Geometric Dimensioning and Tolerancing (GD&T)

– Simplified ASME Y14.5 2009

A Guide to Application and selection
of Geometric Tolerance

BIPINKUMAR SINGH

BlueRose Publishers

First Published in June 2021

ISBN: 978-93-5427-731-3

BLUEROSE PUBLISHERS
www.bluerosepublishers.com
info@bluerosepublishers.com
+91 8882 898 898

Cover Design:
Nirmal

Typographic Design:
Namrata Saini

Distributed by: BlueRose, Amazon, Flipkart, Shopclues

Acknowledgments

Firstly Credit is gratefully given to Society of Mechanical Engineers (ASME) and acknowledgment made for the use of definitions and terms from the ASME Y14.5M-2009 Dimensioning and Tolerancing Standard.

I want thank my family Mom, Dad, My wife Priya and Daughter Krisha for motivating and supporting me throughout. I want to thank Khushi for being helping hand in documentation related things.

Special Thanks to Ankit Agarwal for helping me and putting more examples in this book. Special Thanks to Paresh Wani for supporting me from start.

I want to thanks Hemal Rajput, Chanchalesh Vishwakarma, Ankit Agarwal, Paresh Wani and my team for doing Proof reading and validation of concept.

I want to thank all my YouTube channel subscribers and viewers. You people are motivation for me to go one step further to reach you all.

Thank you!!

Note to Students and Readers

Dear Readers,

Welcome to learning Geometric dimensioning and Tolerancing. This book provides you almost everything regarding GD&T, from concept to applications. I have tried to put everything I used during my 15 years of experience.

During reading you must make an effort to use it successfully in your projects. My faith is about the Journey, journey should be beautiful. A beautiful journey will bring you success. So enjoy the journey. You can easily access and watch my short Videos on these topics available on my website Essomec.com

Learn it step by step, one by one concept. You cannot memories everything thing. You have to read and apply. You have to understand and practice it so it reaches to your subconscious and eventually it will lead you to Geometric dimensioning and tolerancing world.

Contents

References: The Content is based on 15 years of experience. This provides simplified understanding of ASME Y14.5 2009. Book covers almost all concepts and explains in simplified manner.

Assumptions:

Third Angle projection method is used in this Book and Inch Tolerance system has been followed.

Any dimension without a tolerance on drawings views are subjected to general tolerances on drawing.

Chapter 1

Engineering Drawing and GD&T

1.1 Engineering Drawing:

An Engineering drawing is a communication document that has precise detail of part. It includes following details:

-Defines Geometry (shape, size, and form)

-Defines Tolerances

-Material, Heat treatment, Coatings etc.

-Functional Relationship and

-Documentation

Today CAD tools e.g. Solidworks, Creo Parametric etc. have made this easier. Initially it was done manually using T Square, Drawing boards, different size of drawing sheets etc.

Whether engineering drawing created manually or by CAD tools, intent remains same. It is a document that communicates designer's requirement to manufacturer.

Failure in communication can result in improper product and cost. To avoid this clear communication is needed.

Any Drawing error will cost lot more if product is manufactured and delivered to the customers. Drawing error may cost: Time, Money, Material or Unhappy Customer. Our Focus should be precise communication related to information of product example coating, Hardfacing etc. Apart from this Dimensioning and Tolerancing are important Factors We will be discussing in this Chapter.

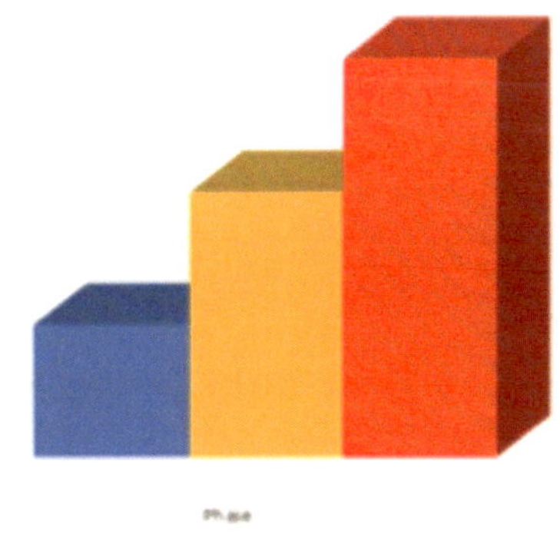

Figure 1.1 Cost Vs Phase

1.2 Introduction to Dimensioning and Tolerancing:

Dimension:

Numerical value used to define the form, size, orientation or location of a part. It is represented in appropriate units of measure.

Tolerance:

The total amount or the extreme limits within which dimension allowed to vary. The Tolerance zone is difference between upper and lower limit of dimension.

Bilateral:

A tolerance in which dimension varies in both directions from the specified value. It can be equal bilateral that is also called Plus minus Tolerance or Unequal Bilateral as shown below:

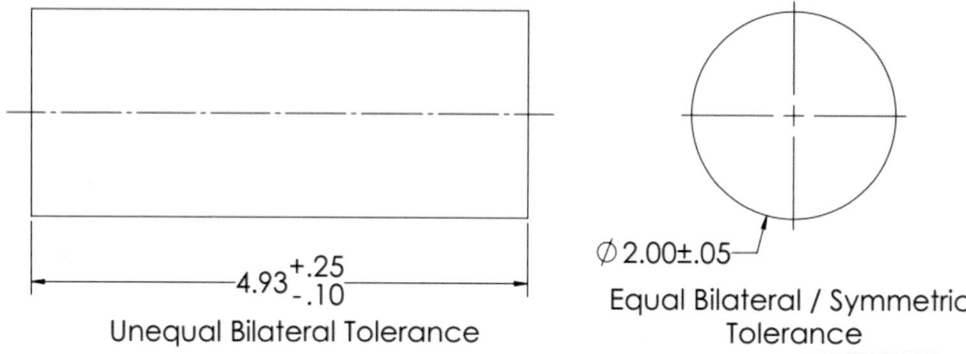

Figure 1.2

Geometric Tolerance:

The tolerances used to control size, form, profile, orientation, location and runout. For example:

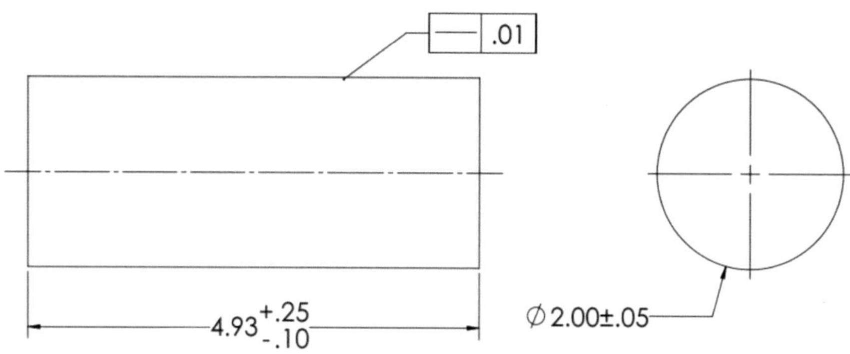

Figure 1.3

Unilateral and Limit:

A Unilateral tolerance is in which Dimension variation is allowed either in one direction from the given dimension value.

A Limit tolerance is when low and high limit of dimension mentioned, with high value on top and low value on bottom.

For example:

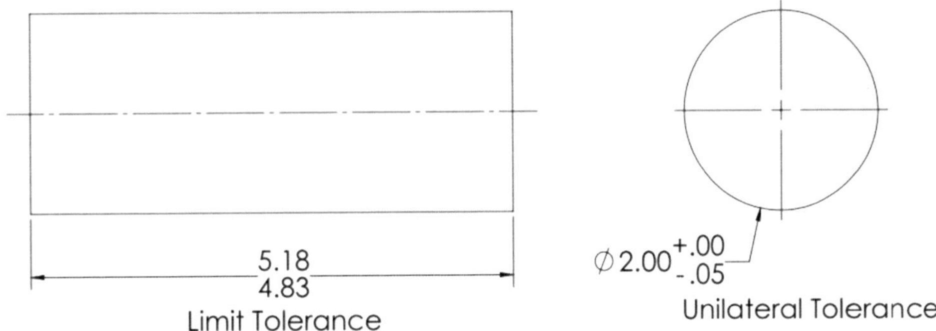

Figure 1.4

1.3 Fundamental Dimensioning Rules

Engineering Intent must be clearly specified in drawing by dimensioning and tolerancing. Following are the fundamental rules of dimensioning:

(a) Each dimension on drawing must have a tolerance, except for those dimensions specifically identified as reference, maximum, minimum or stock (commercial stock size).

(b) Dimensioning and tolerancing shall be complete for part or assembly so there are full understandings of each features (geometry and shape).

(c) Use of Reference dimensions should be minimized. Each necessary dimension must be shown for a part or feature. No extra dimensions than necessary for complete definition of parts shall be given.

(d) Selected Dimensioning scheme and arrangement (chain dimensioning, baseline dimensioning etc.) to suit the function and mating relationship of a part and shall not have more than one interpretation.

(e) The drawing should define part geometry without specifying manufacturing methods. However, in those instances where manufacturing, processing, and other information is necessary to the definition of engineering requirements, it shall be specified on the drawing or in a procedure document mentioned as a note on the drawing.

(f) Certain processing dimensions that provide for finish allowance, shrink allowance, and other requirements, provided the final dimensions are given on the drawing. Nonmandatory processing dimensions shall be identified by an appropriate note, such as NONMANDATORY.

(g) Dimensions should be arranged to provide required information for optimum readability. Dimensions should be shown in true profile views and refer to visible outlines.

(h) Cables, wires, rods and other materials manufactured to gage or code numbers shall be specified by linear dimensions indicating the diameter or thickness. Gage or code numbers may be shown in parentheses following the dimension.

(i) A 90° angle and 0° implied angles apply where center lines and lines depicting features are shown on a drawing at right angles or parallel and no angle is specified.

(j) A 90° basic and 0° basic implied angles apply where center lines of features in a pattern or surfaces shown at right angles or parallel on the drawing are located or defined by basic dimensions and no angle is specified.

(k) Unless otherwise specified, all dimensions are applicable at 20°C (68°F). Compensation may be made for measurements made at other temperatures.

(l) All dimensions and tolerances apply in a Free State condition. This principle does not apply to non rigid parts.

(m) Unless otherwise specified, all geometric tolerances apply for full depth, length, and width of the feature.

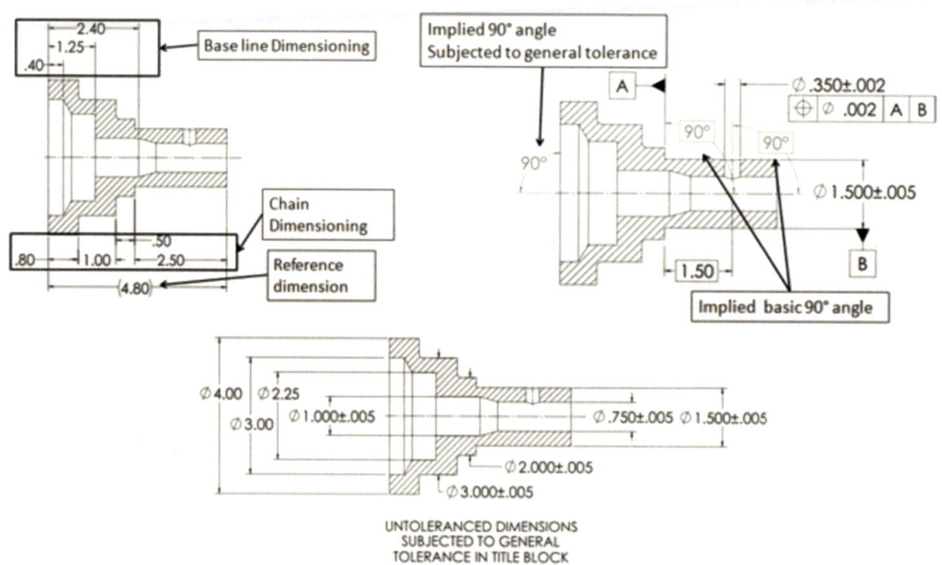

Figure 1.5

These rules shall be followed for clear intent of drawing.

1.4   Co-Ordinate Tolerancing System:

More than hundred years coordinate tolerancing was the predominant tolerancing system used on engineering drawings. Coordinate tolerancing is a method of dimensioning of a part feature is located or defined by means of rectangular dimensions with given tolerances. For Example:

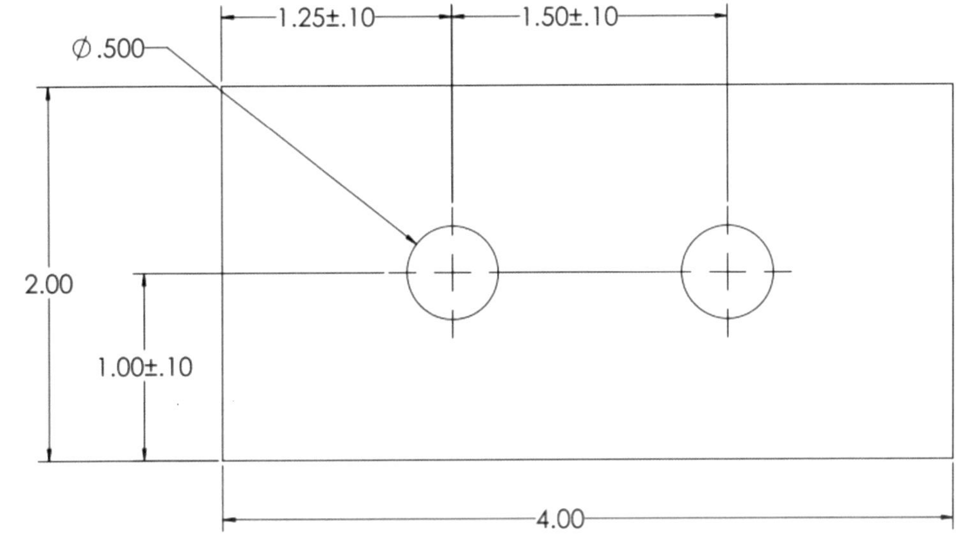

Figure 1.6

This Method will result in a Square or rectangular Tolerance zone. The tolerance zone is fixed for manufacturer, even acceptable parts can be rejected, and ambiguous inspection method as shown in figure 1.7.

The Tolerance zone ±.10 results in position shift of .28 diagonally but .2 in vertical or horizontal direction. So, it is not uniform throughout. The inspector can place part in two ways as shown in figure 1.7. Both measurements can be different and good part may be rejected or bad part is selected. This is due to no clear instruction for measurement. These shortcomings will be eliminated by Geometric tolerancing. This type of dimensioning good to use for size but not for controlling Orientation, Location or Form.

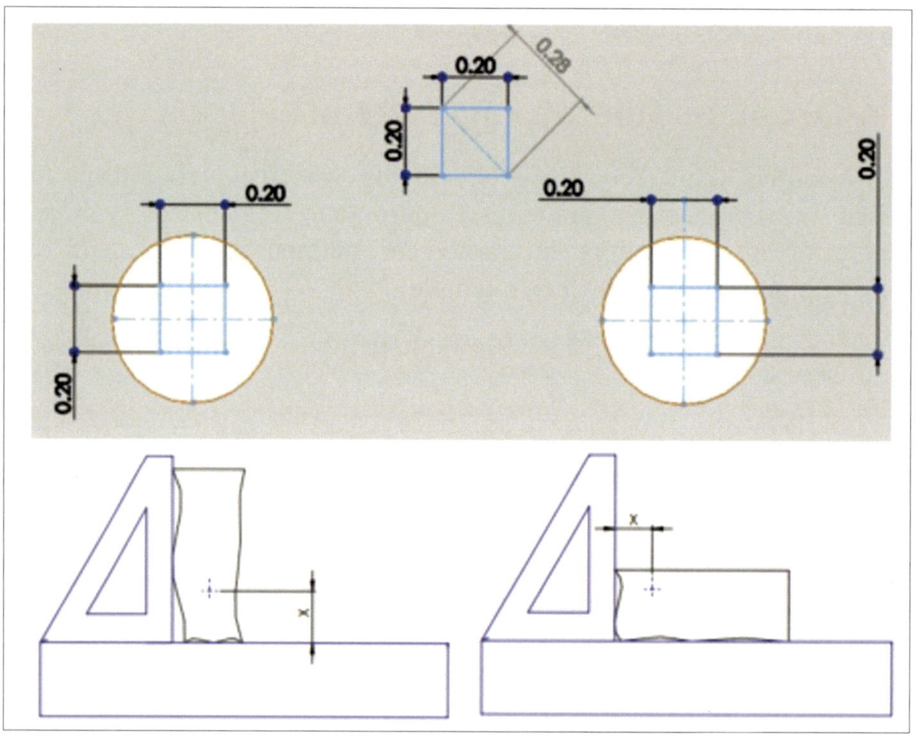

Figure 1.7

1.5 Introduction to Geometric Dimensioning and Tolerancing (GD&T)

Geometric Dimensioning and Tolerancing is a language, used on drawing for accurately defining a part. We can say it is a precise way to define size, form, Orientation and Location of part and Features.

For all precise communication, words are usually inadequate. For example, a note on the drawing saying, THE PART SHOULD HAVE GOOD STRAIGHTNESS. This note on drawing does not convey proper control WHAT IS A GOOD STRAIGHTNESS.

During Second World War, a specialized language based on graphical representations and math has evolved to improve communication. This was due to rapid production and assembly failure. It was found that clear intent was never communicated. In its current form, the language is recognized throughout the world as *Geometric Dimensioning and Tolerancing (GD&T).*

Reference for GD&T

The following American National Standards define GD&T's vocabulary and provide its grammatical rules.

- ASME Y14.5M-2009, Dimensioning and Tolerancing. It is revision and update of ASME Y14.5-1994.

GD&T USE :

GD&T provides uniformity in drawing and its clear intent, thereby reducing miscommunication, measurement from drawing or guesswork, and any shape or size assumptions. Every Department of a company works in simulated manner.

The use of GD&T can improve your product by providing designers a tool for clear communication by following the functional dimensioning philosophy.

There are two ways tolerances are increased using GD&T. First, it makes tolerance zone cylindrical so provides uniformity and second under certain conditions, GD&T provides Bonus tolerance for manufacturing. This additional tolerance can make a significant savings in production costs.

Using functional dimensioning, the tolerances are assigned to the part based upon its functional requirements. This often results in a larger tolerance for manufacturing.

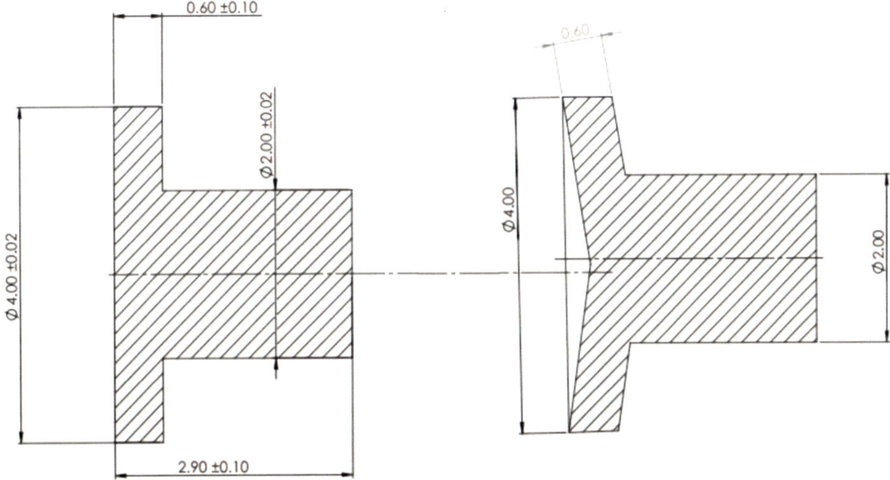

Figure 1.8: Drawing that does not use GD&T

In Fig.-1.8, we have not applied any Geometric Tolerancing. We can see that there is no orientation and Coaxiality control. Unusable part can be accepted in inspection and it will fail in assembly or in function. Same part can be given with Orientation and Coaxiality control as shown in figure 1.9. so that manufactured part can fit in assembly and functionality both.

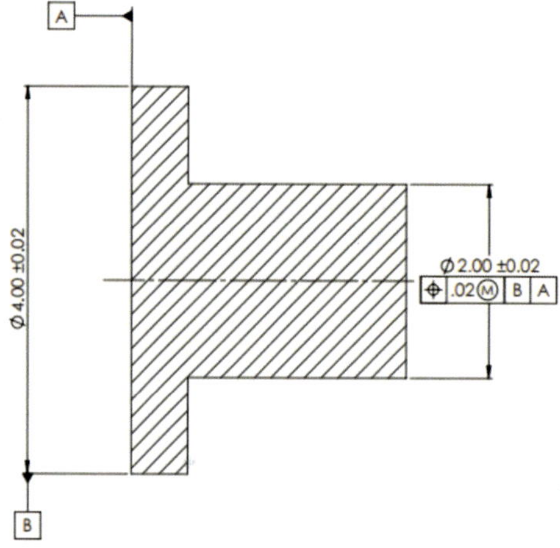

Figure 1.9: Drawing with GD&T

GD&T eliminates drawback of coordinate dimensioning by providing cylindrical tolerance zone. And by using MMC modifiers, we can get Bonus tolerance or extra tolerance we will discuss that in detail in following chapters. By using Datums we provide base for inspection and Datum Reference Frame.

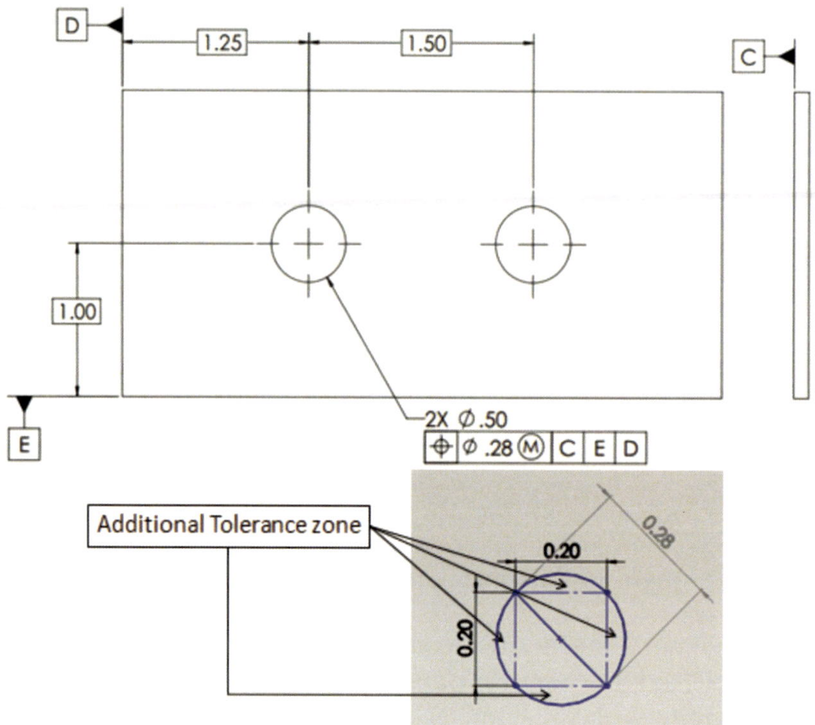

Figure 1.10: Drawing that uses GD&T

1.10. This also removes ambiguity in Inspection by specifying Primary Secondary and Tertiary Datum. By using positional tolerance of Ø.28 we get additional tolerance as shown in figure. It increases tolerance by 57%.

1.6 Myths In GD&T

a) <u>GD&T increases the cost.</u> The Fact is, it saves MONEY if properly used.

b) <u>Dimensioning and Tolerancing are separate step</u>. In fact, method of dimensioning and tolerancing must be decided based on **functionality.**

c) <u>Geometric Tolerance should be given to Critical part and Feature</u>. It can be used for simple component to control size, form orientation or location.

d) <u>Application of Geometric Tolerance takes much time.</u> With Understanding and Practice and CAD tools it becomes easier to use international language of Drawing.

e) <u>It Improves Design.</u> It is a precise communication between designer's intent and Production shop for understanding but it does not improve design. If Part is designed using any wrong methodology or component cannot match functional requirement, geometric dimensioning and tolerancing cannot correct that.

Chapter 2

Definition and Symbols

2.1 Definitions

We should use vocabulary as ASME Y14.5 2009 does. In this section we will see various terminologies used in ASME Y14.5 2009

Part Features

Usually, a part feature is a single surface (or a pair of opposed parallel plane surfaces) having uniform shape.

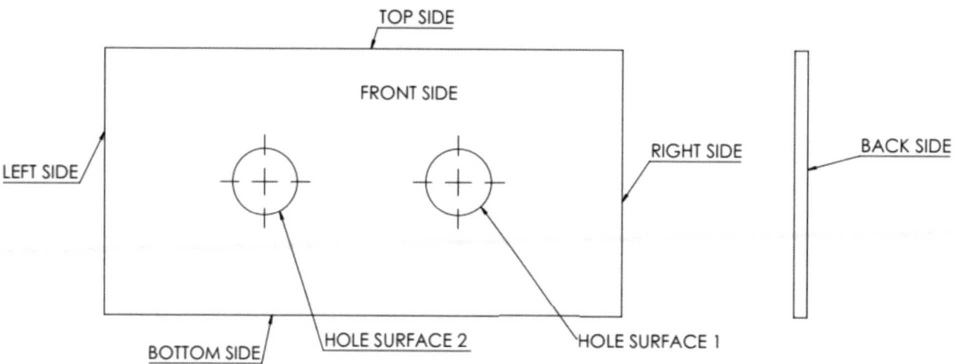

Figure 2.1: Part Feature

ALL SURFACES OF THE PART CONSIDERED AS FEATURE

Features of Size

A *feature of size* is one cylindrical or spherical surface, or a set of two opposed elements or opposed parallel surfaces, associated with a size dimension.

A feature of size has opposing points that partly or completely enclose a space, giving the feature an intrinsic dimension—size—that can be measured apart from other features. Features of size are subject to the principles of material condition modifiers.

Internal, External and Non-Feature of Size:

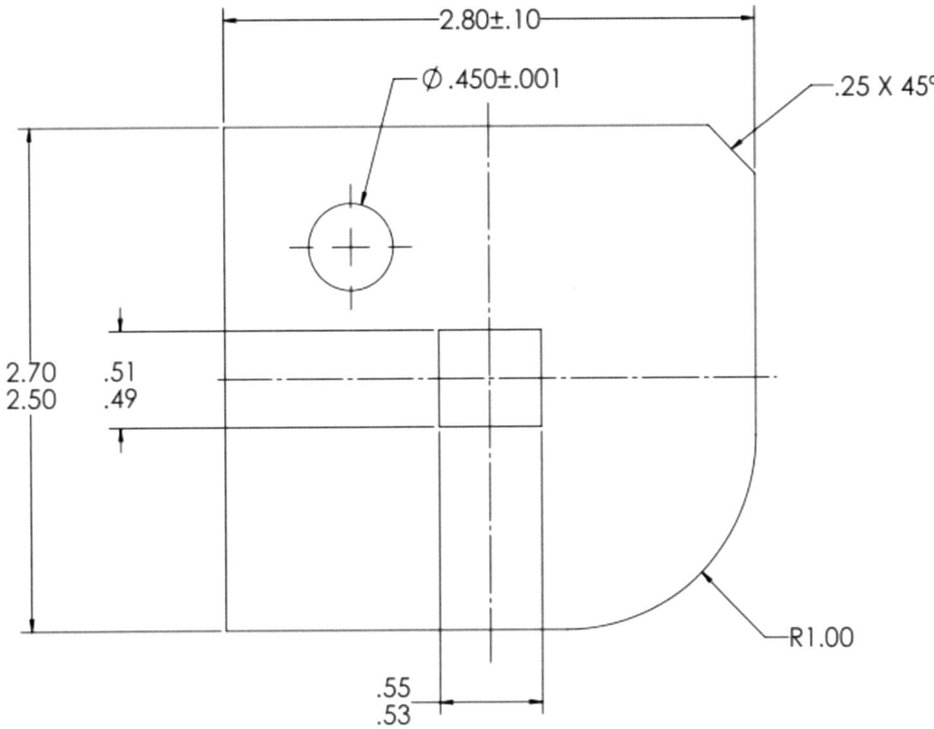

Figure 2.2: Part Feature Internal, External and Non-Feature of size

Figure 2.2 show 2.80±.10 and 2.50-2.70 as external feature of size. Diameter 4.500±.001, .51-.49 and .55-.53 as internal feature of size. Fillet and Chamfer are non-feature of size.

A *non-size feature* is a surface having no unique or intrinsic size (diameter or width) dimension to measure. Non size features include the following:

- A nominally flat planar surface
- An irregular or "warped" planar surface, such as the face of a windshield.
- A *radius*—a portion of a cylindrical surface encompassing less than 180° of arc length.
- A *spherical radius*—a portion of a spherical surface encompassing less than 180° of arc length.
- A *revolute*—a surface, such as a cone, generated by revolving a spine about an axis.

Irregular Feature of size

The type of feature that is neither a sphere, cylinder, nor width-type feature, yet clearly has "a set of two opposed elements."

The D-hole or any other shape is called an "irregular feature of size" by some drafting manuals, Irregular feature of size are not much covered in ASME Y14.5 2009.

Actual Mating Envelope:

Actual mating envelope is variable value derived from manufactured part. Smallest size that can be contracted to an external feature OR Largest size that can be expanded to an internal feature is call as Actual mating Envelope. See Figure 2.3

Unrelated actual mating Envelope:

Unrelated actual mating Envelope is the Max size that can be expanded for internal feature and smallest size that can be contracted to external Feature Irrespective of any Datum Reference.

Related actual mating Envelope:

Related actual mating Envelope is the Max size that can be expanded for internal feature and smallest size that can be contracted to external Feature and oriented with respect to Datums. Figure 2.3 shows the Related Actual mating envelope is oriented w.r.t. datum A

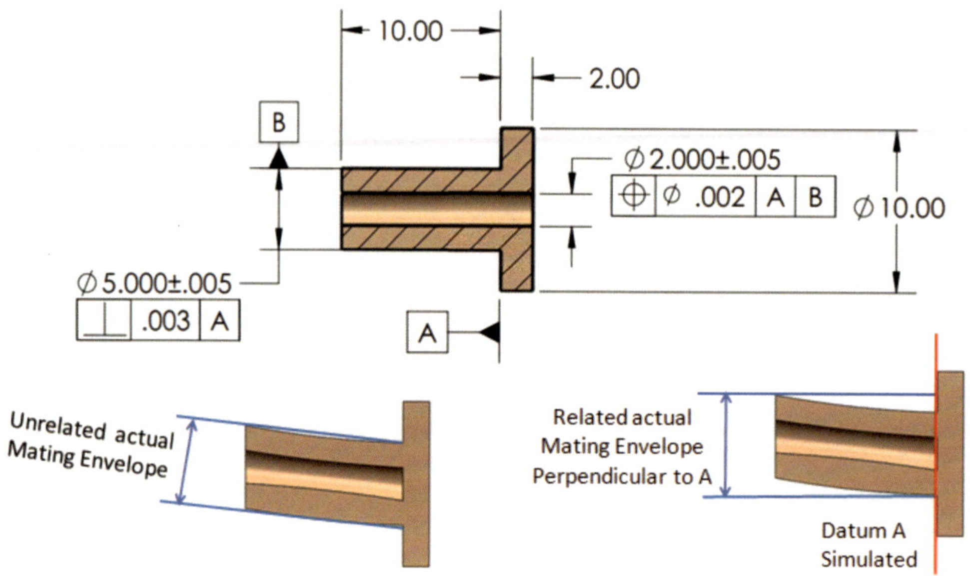

Figure 2.3: Actual Mating Envelope

Related and Unrelated actual minimum mating Envelope:

Related and unrelated material envelope is inside the material.

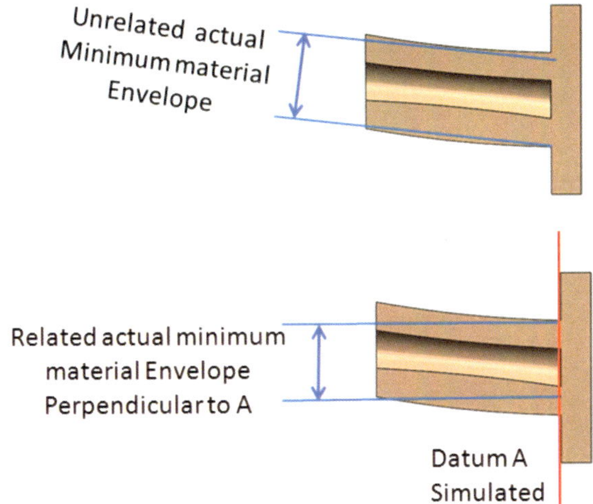

Figure 2.4: Actual Mating Envelope

Maximum material Condition (MMC):

When part has maximum material, it is called as Maximum material Condition MMC. For example for a hole it is smallest diameter of feature of size and for shaft the largest Diameter of the feature of size is its MMC.

Least material Condition (LMC):

When part has minimum material, it is called as least material Condition LMC. For example for a shaft it is smallest diameter of feature of size and for hole It is Largest diameter of feature of size is its LMC.

Regardless of Feature Size (RFS):

Regardless of feature size is a condition that Geometric Tolerance is applicable at any increment of manufactured size within Limit of size.

Inner Boundary(IB):

Boundary Created by smallest feature minus stated geometric tolerance and any other tolerance resulting due to departure of feature from its stated material condition (Bonus tolerance, will be discussed in detail in subsequent chapters).

Outer Boundary (OB):

Boundary Created by largest feature plus stated geometric tolerance and any other tolerance resulting due to departure of feature from its stated material condition (Bonus tolerance, will be discussed in detail in subsequent chapters).

Least Material Boundary (LMB):

Boundary That Exists on or inside of the material defined by cumulative effect of tolerance of datum of precedence. Term is discussed in detail subsequent chapters.

Maximum Material Boundary (MMB):

Boundary That Exists on or outside of the material defined by cumulative effect of tolerance of datum of precedence. Term is discussed in detail subsequent chapters.

Datum Feature Simulator:

Datum feature simulator is inverse shape of the datum feature. It is used to establish datum from Datum Feature. In ASME 2009 Y14.5 it is mainly theoretical simulator. Physical Datum simulators are Surface plates, Mandrels, gages etc.

Basic Dimension:

It is theoretically exact dimension. Any dimension made basic dimension must have some indirect tolerance from Geometric Tolerance or by any other mean e.g. tolerance note on drawing such as ALL HOLE LOCATION DEFIND BY BASIC DIMENSION FOLLOWS $\boxed{\oplus\ |\ \varnothing\ .002\ \text{Ⓜ}\ |\ A\ |\ B\ |\ C}$.

Feature Relating Tolerance Zone Framework (FRTZF):

The Tolerance zone framework that constrains rotational degree for freedom relative to datums for features in pattern is FRTZF. We will discuss in detail with Composite tolerancing.

Pattern Locating Tolerance Zone Framework (PLTZF):

The Tolerance zone framework that constrains translational and rotational degree for freedom relative to datums for features in pattern is FRTZF. We will discuss in detail with Composite tolerancing.

Regardless of material Boundary (RMB):

In this datum simulator size is variable, it originates at MMB and moves towards LMB and makes maximum contact with datum feature to establish datum.

Radius and Controlled radius:

Difference between Radius and Controlled Radius as Shown:

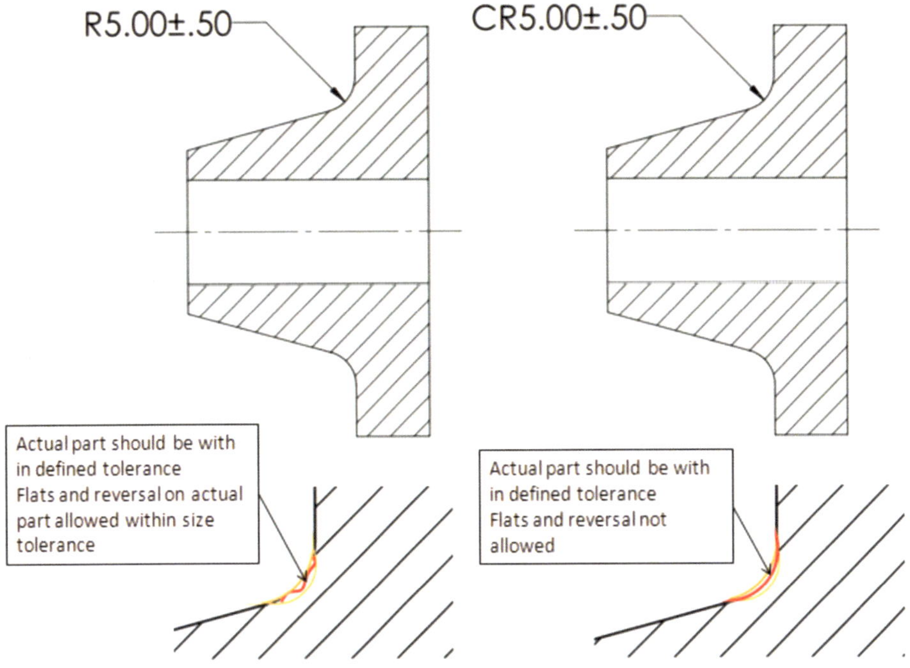

Figure 2.5 Controlled Radii

2.2 Symbols

Feature Control Frame:

Each geometric control for a feature is conveyed on the drawing by a rectangular sign called a *feature control frame*.

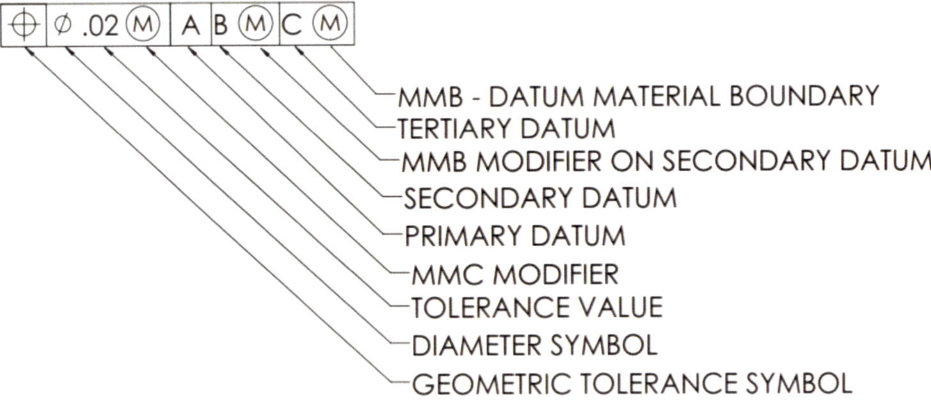

Figure 2.4: Compartments that make up the feature control frame.

- The first section of frame shows *Geometric Tolerance Symbol.*
- The second section of feature control frame is for Tolerance and related modifiers and symbols as shown in figure 2.4.
- The other section shows datums in order of precedence. This also indicates MMB or LMB symbols if applicable.

Geometric Characteristic Symbols:

Symbol	Characteristic	If only RFS	Tolerance Type	Application and Datum Requirement
—	Straightness		Form	Individual Feature (No Datum)
▱	Flatness			
○	Circularity	RFS		
⌭	Cylindricity	RFS		
⌒	Profile of a Line	RFS	Profile	Individual and Related (With Datum, Without Datum)
⌓	Profile of a Surface	RFS		
∠	Angularity		Orientation	Related Control (Datum Needed)
⊥	Perpendicularity			
∥	Parallelism			
⊕	Position		Location	May be related or Unrelated
◎	Concentricity	RFS	Location	Related Control

	Symmetry	RFS		(Datum Needed)
	Circular Runout	RFS	Runout	Related Control (Datum Needed)
	Total Runout	RFS		

Table 1: Symbols

Table 1 shows all characteristic symbols of Geometric Tolerance. It also shows additional information for few symbols applicable on RFS basis only. We will discuss this rule#2 in detail.

Modifier Symbols:

Symbol	Terminology
M	Maximum Material Condition MMC
L	Least Material Condition LMC
P	Projected Tolerance Zone
F	Free State
T	Tangent Plane
U	Unequally disposed Profile Tolerance

⬡ S T	**Statistical Tolerance Application**
▷	**Translation**
⌀	**Diameter**
S⌀	**Spherical Diameter**
R	**Radius**
SR	**Spherical Radius**
CR	**Controlled Radius**
( )	**Reference**
A ◀──▶ B	**Between**
⟲	**All around**
⊙	**All Over**

Table 2: Symbols

In Subsequent chapters we will discuss these modifiers as we use them.

There are other symbols as well, this text explains most used ones.

Chapter 3

Rules and Different concepts of GD&T

3.1 Limit of Size:

Limit of size must be followed; Actual local size at each cross section of the feature or part must be with in size tolerance. Its size and form can vary with in size limit unless specified.

3.2 Rule#1 –

Form of individual feature is controlled by limit of size tolerance.

At MMC of feature of size, no form variation is permitted i.e. perfect form is needed at MMC.

We need perfect form at MMC unless Straightness or Flatness is mentioned on size dimension or Independency symbol is used.

Where actual manufactured product size departs from MMC, for variation is permitted. The variation permitted is the amount of departure at MMC.

When feature is produced at LMC, we get maximum form variation as perfect form at LMC no needed.

In case if LMC modifier is used we need Perfect form at LMC.

Rule#1 is individual feature rule, and this ensures the easy assembly as per designer's intent.

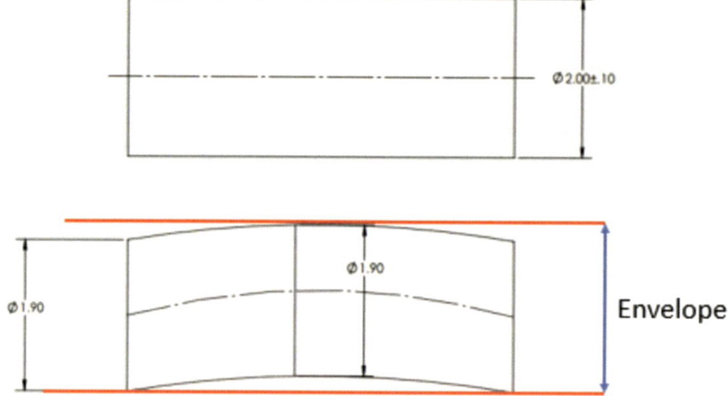

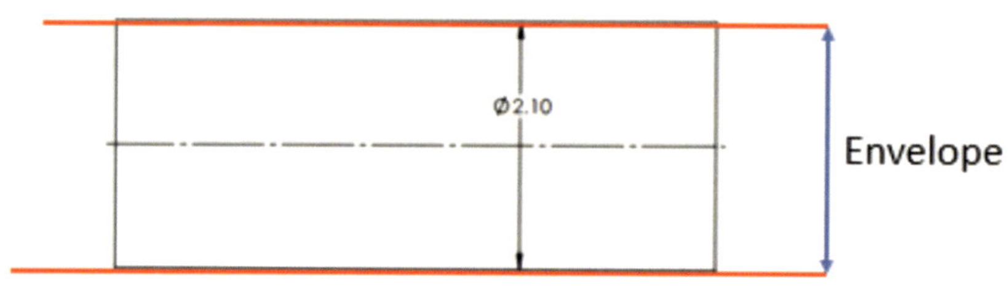

Envelope = MMC = VC = 2.1
Part manufactured at size below to MMC gets Form tolerance of amount of departure
If it is manufactured at LMC as shown gets max Form Tolerance = MMC-LMC

Figure 3.1

Envelope = MMC = VC = 2.1
Perfect form needed when manufactured at MMC

Figure 3.2

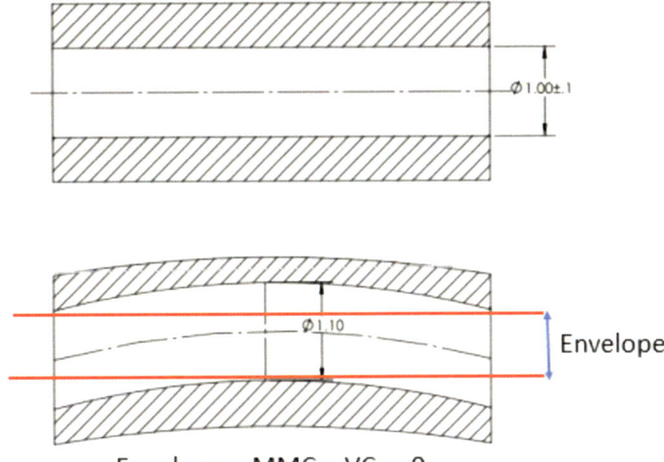

Figure 3.3

Rule#1 does not apply to stock such as tube, bars Plates etc. as those governed by industry or government standard. And Part made from those carries same standard until any Geometric Tolerance is mentioned. This (**rule#1**) also do not apply where parts subjected to Free State variation and when unrestrained.

We can override Rule #1 if independency symbol or Geometric Tolerance (straightness, Flatness) Applied to size dimension.

3.3 Rule#2 –

No Modifiers are used to denote RFS or RMB condition. MMC, LMC, MMB, or LMB shall be specified on the drawing where it is required. If nothing is mentioned, we assume RFS.

Sometime Form deviations are so extreme that Tolerance in terms of axis does in equivalent to the tolerance in terms of surface (Virtual Condition). In those scenarios, Tolerance in terms of surface takes precedence. We will discuss this in detail in Positional tolerance section.

Circular Runout, total Runout, concentricity, profile of a line, profile of a surface, circularity, Cylindricity, and symmetry tolerances are applicable only on an RFS basis and cannot be modified to MMC or LMC. It is shown in TABLE 1.

3.4 Virtual Condition (VC):

The constant boundary generated by collective effect of Geometric Tolerance and size. This boundary shall not be violated in any scenario.

Feature frame with MMC modifier:

VC internal Feature = MMC-Geometric Tolerance

VC External Feature = MMC + Geometric Tolerance

Feature frame with LMC modifier:

VC internal Feature = LMC + Geometric Tolerance

VC External Feature = LMC – Geometric Tolerance

Multiple virtual condition may be possible if multiple Feature tolerance frame with different tolerance applied. Each virtual condition must be satisfied.

3.5 Resultant Condition (RC):

Single Worst case boundary generated LMC or MMC size. It is collective effect of Geometric Tolerance, Bonus tolerance and Size.

Feature frame with MMC modifier:

RC internal Feature = LMC + Geometric Tolerance + Bonus

RC External Feature = LMC – Geometric Tolerance - Bonus

Feature frame with LMC modifier:

RC internal Feature = MMC - Geometric Tolerance - Bonus

RC External Feature = MMC + Geometric Tolerance + Bonus

3.6 Effect of RFS:

Wherever RFS gets applied with Geometric Tolerance, the tolerance value (zone) is applicable to all manufactured size within size limit. The tolerance gets applied regardless of size of unrelated actual mating envelope. Inner Boundary IB and Outer boundary can be calculated for RFS as shown in figure 3.4 for Internal Feature and 3.5 for External Feature.

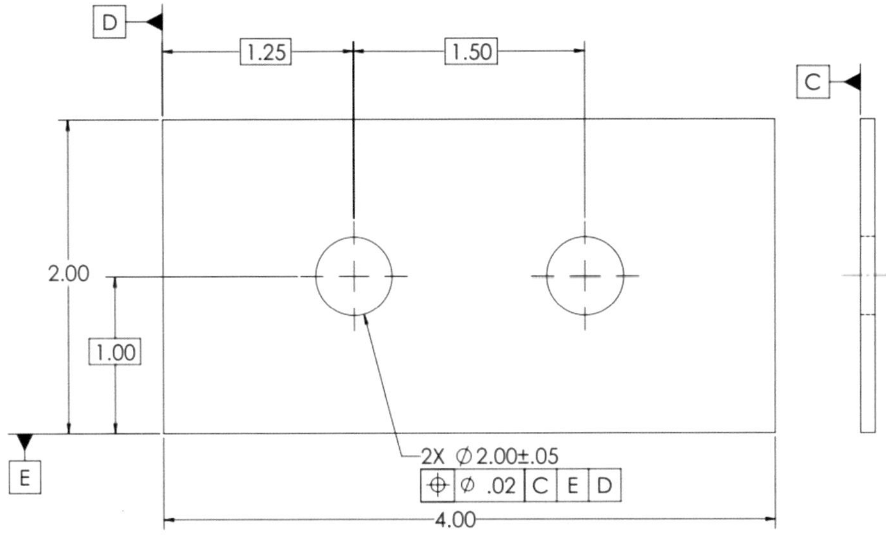

Figure 3.4 – Internal Feature at RFS

Hole Diameter	Positional Tolerance	IB	OB
MMC = 1.95		1.93	
1.96			
1.97	.02		
2			
2.01			
LMC = 2.05			2.07

Table 3.1 (Calculation of IB and OB at RFS for Internal feature)

For Internal feature Worst case boundary = IB = MMC – Geometric Tolerance

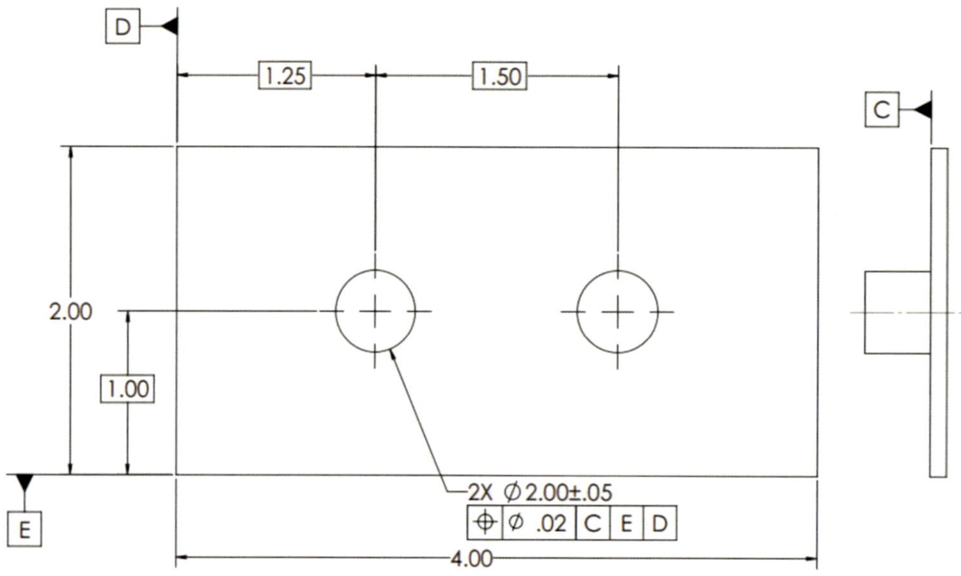

Figure 3.5- External feature at RFS

Hole Diameter	Positional Tolerance	IB	OB
LMC = 1.95		1.93	
1.96			
1.97			
2	.02		
2.01			
MMC = 2.05			2.07

Table 3.2 (Calculation of IB and OB at RFS for external feature)

For External feature Worst case boundary = OB = MMC + Geometric Tolerance

3.7 Effect of MMC:

Whenever MMC modifier is applied with Geometric Tolerance, the specified tolerance value (zone) is applicable to specific size of unrelated actual mating envelope i.e. at MMC.

If size of unrelated actual mating envelope departs from MMC we get some additional bonus tolerance. The amount of bonus tolerance is equivalent to departure amount. Max bonus we get if feature is at LMC. See Figure 3.6

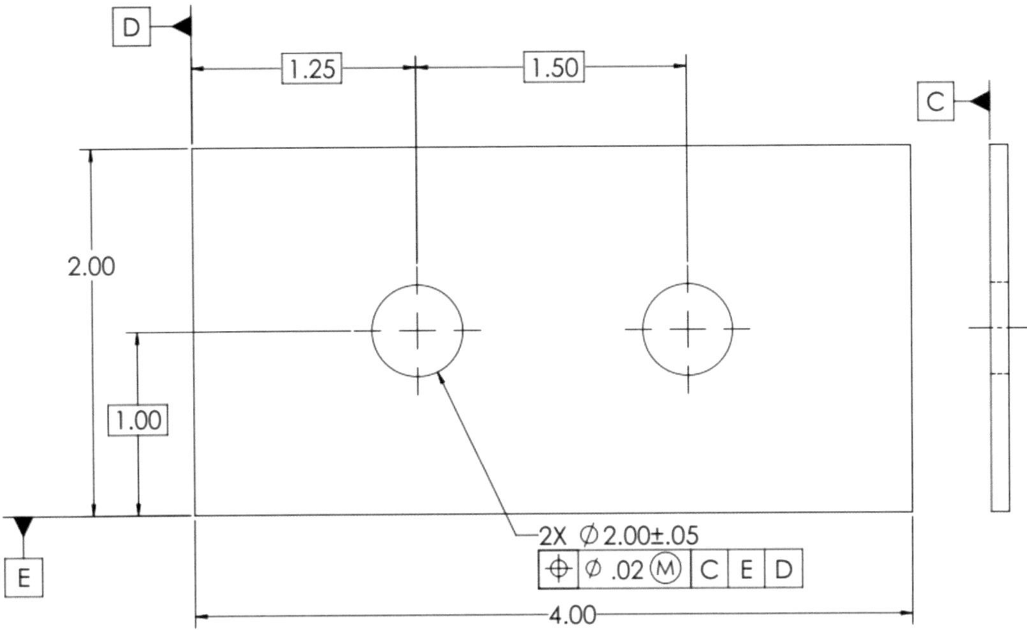

Figure 3.6: MMC to internal Feature

Hole Diameter	Positional Tolerance	VC	RC
MMC = 1.95	.02		
1.96	.03		
1.97	.04	1.93	
2	.07		
2.01	.08		
LMC = 2.05	.12		2.17

Table 3.3 (Calculation of VC and RC with mmc Modifier for Internal feature)
Virtual Condition = MMC – Geometric Tolerance
Resultant boundary = LMC + Geometric Tolerance + Bonus

Table 3.3 shows effect of MMC on internal feature. We get .02 tolerance for 1.95 size manufactured and if manufactured at LMC we get .12 tolerance. Virtual condition (VC) is a constant boundary 1.93 for all manufactured size. Resultant condition is a single value for internal feature it is LMC + Geometric Tolerance + Bonus.

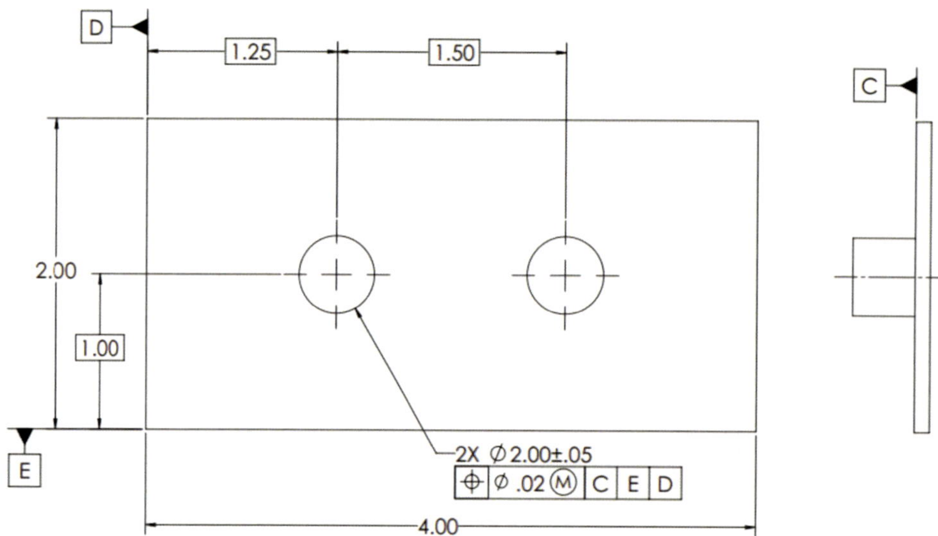

Figure 3.7: MMC to external feature

Hole Diameter	Positional Tolerance	VC	RC
LMC = 1.95	.12		1.83
1.96	.11		
1.97	.10		
2.03	.04	2.07	
2.04	.03		
MMC = 2.05	.02		

Table 3.4 (Calculation of VC and RC with mmc Modifier for external feature
Virtual Condition = MMC + Geometric Tolerance
Resultant boundary = LMC - Geometric Tolerance – Bonus

3.8 Effect of LMC:

Wherever LMC modifier is applied with Geometric Tolerance, the specified tolerance value (zone) is applicable to one size of unrelated actual mating envelope i.e., at LMC.

If size of unrelated actual mating envelope departs from LMC we get some additional bonus tolerance. The amount of bonus tolerance is equivalent to departure amount. Max bonus we get if feature is at MMC. See Figure 3.8.

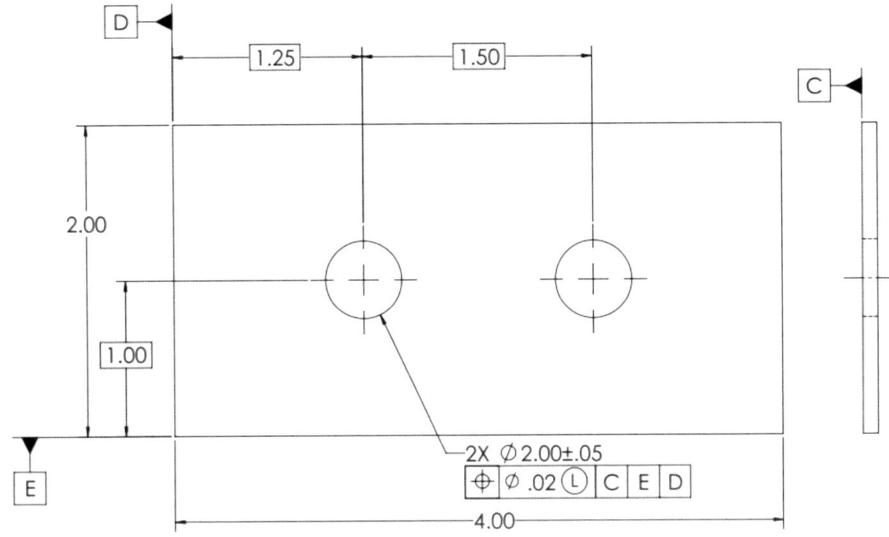

Figure 3.8: LMC to internal Feature

Hole Diameter	Positional Tolerance	VC	RC
MMC = 1.95	.12		1.83
1.96	.11		
1.97	.10	2.07	
2	.07		
2.01	.06		
LMC = 2.05	.02		

Table 3.5 Calculation of VC and RC with LMC Modifier for internal feature)
Virtual Condition = LMC + Geometric Tolerance
Resultant boundary = MMC - Geometric Tolerance - Bonus

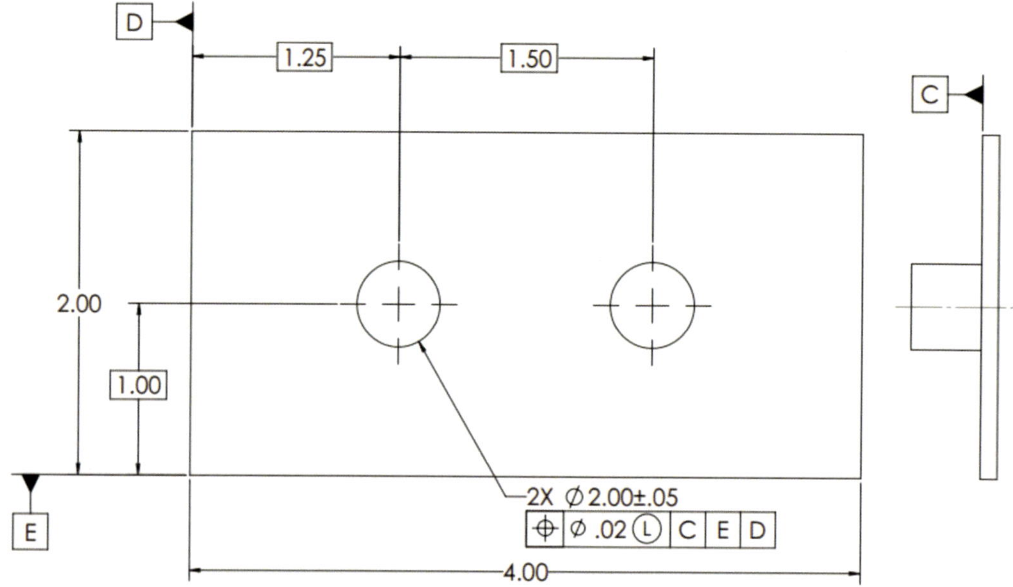

Figure 3.9: LMC to external feature

Hole Diameter	Positional Tolerance	VC	RC
LMC = 1.95	.02		
1.96	.03		
1.97	.04	1.93	
2.03	.10		
2.04	.11		
MMC = 2.05	.12		2.17

Table 3.6 Calculation of VC and RC with LMC Modifier for external feature)
Virtual Condition = LMC - Geometric Tolerance
Resultant boundary = MMC + Geometric Tolerance + Bonus

3.9 Bonus Tolerance:

As we have already discussed, we get bonus tolerance when MMC or LMC modifiers are in feature control from.

If Modifier is MMC and part produced at LMC we get maximum bonus.

If Modifier is LMC and part produced at MMC we get maximum bonus.

3.10 Comparison Table:

Boundaries	Internal Feature @ RFS	External Feature @ RFS
IB	MMC – Geometric Tolerance	LMC- Geometric Tolerance
OB	LMC + Geometric Tolerance	MMC + Geometric Tolerance

Boundaries	Internal Feature @ MMC modifier	External Feature @ MMC Modifier
VC	MMC - Geometric Tolerance	MMC + Geometric Tolerance
RC	LMC + Geometric Tolerance + Bonus	LMC - Geometric Tolerance- Bonus
Bonus	LMC-MMC	MMC – LMC

Boundaries	Internal Feature @ LMC modifier	External Feature @ LMC modifier
VC	LMC + Geometric Tolerance	LMC - Geometric Tolerance
RC	MMC - Geometric Tolerance- Bonus	MMC + Geometric Tolerance + Bonus
Bonus	LMC-MMC	MMC – LMC

Chapter 4

Tolerance Selection

4.1 Tolerance Selection:

Before discussing Geometric Tolerance and their significance, let us discuss if there is any way to decide which tolerance used at a particular case.

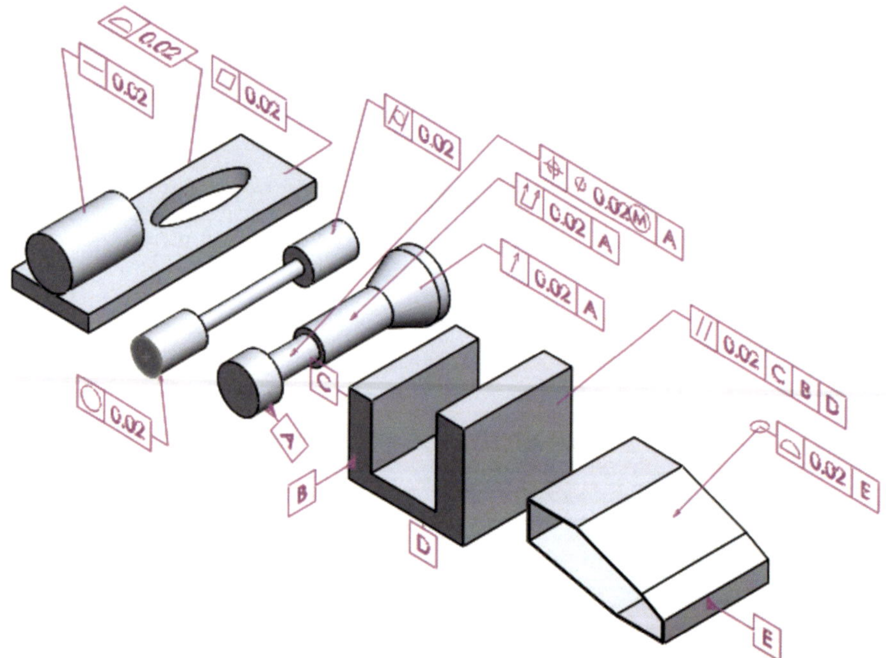

Figure 4.1: Application of Geometric Tolerance

1. Understand functional and assembly requirement.
2. If Feature needs individual control, Form or Profile Tolerance can be used. Based on shape and geometry straightness, flatness, circularity, cylindricity or profile tolerance can be used as shown in Figure 4.1.

3. If Features are related, Establish Datum and select Location, Orientation, Runout, or profile tolerance may be used.

4.2 Tolerance selection (Functional Assembly 1):

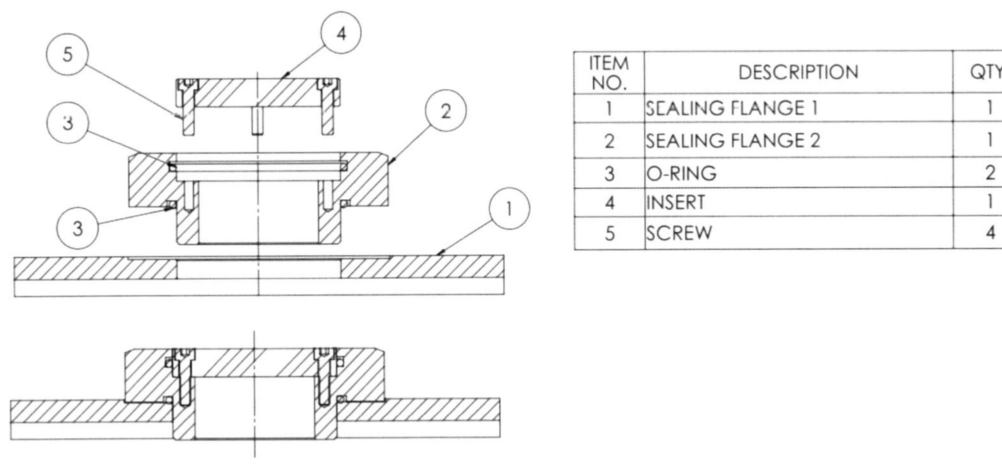

ITEM NO.	DESCRIPTION	QTY.
1	SEALING FLANGE 1	1
2	SEALING FLANGE 2	1
3	O-RING	2
4	INSERT	1
5	SCREW	4

Figure 4.2: Functional assembly 1

The assembly shows 5 components. There are two sealing surfaces, one is flat surfaces (Item 1 and Item 2) and other is surface of revolution (Item 2 and Item 4). The selection can be done based on functional requirement.

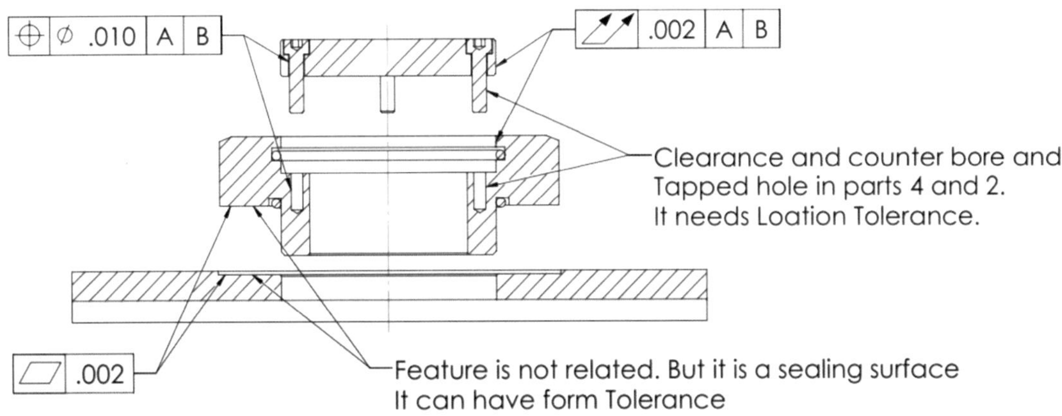

Figure 4.3: Tolerance Selection

1) The flat sealing surfaces (Item 1 and Item 2) can have flatness of form tolerance. It can be refinement of size tolerance. It does not need not to be related to any datum. Instead of Flatness, profile tolerance can be used as it controls size, form,

orientation and location i.e. all in one. Total Runout also can be used to control wobbling and perpendicularity.

2) There are four counter bore holes on item 4 and tapped holes on item 2. Position tolerance related to datums (Axis and Flat surface) to control perpendicularity and location can be used. As there is pattern of size, composite tolerance can be used as well.

3) Runout tolerance on sealing surface of Item 4 and Item 2 can be used.

The selection of tolerances can be always debatable as same surface can be controlled in multiple ways.

4.3 Tolerance selection (Functional Assembly 2):

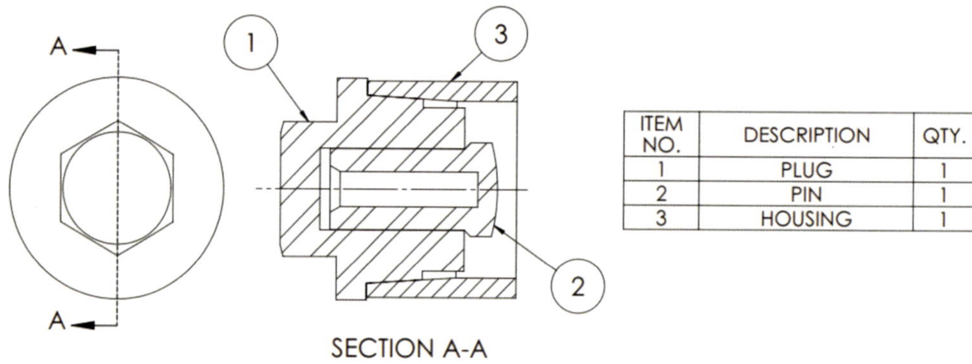

ITEM NO.	DESCRIPTION	QTY.
1	PLUG	1
2	PIN	1
3	HOUSING	1

Figure 4.4: Functional assembly 2

Functional assembly shows item 2 outer diameter and item 1 inner diameter can be controlled irrespective of datum. The taper on plug has another mating surface on housing should be controlled as well.

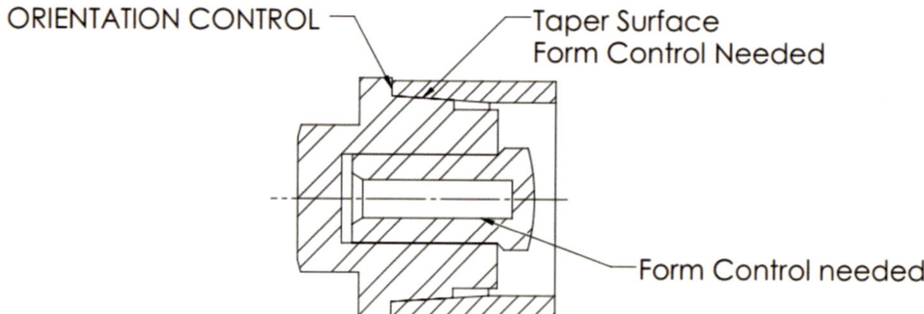

Figure 4.5: Tolerance Selection

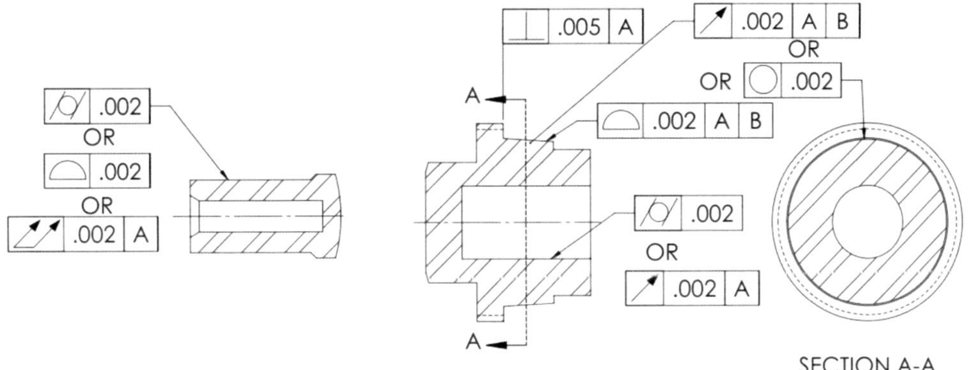

Figure 4.6: Tolerance Selection

1) Item 2 can have Cylindricity control on outer diameter. That would suffice the requirement though Profile of surface and total runout can be used to control the feature of size.

2) As the Item 1 and Item 3 have shoulder contact, orientation tolerance (Perpendicularity) can be used to control that and to be used to establish surface as datum.

3) The Taper surface can be controlled with circular runout. Though profile and circularity tolerance can be used as well.

We will discuss each geometric tolerance in subsequent chapters in detail.

Chapter 5

Form Tolerances

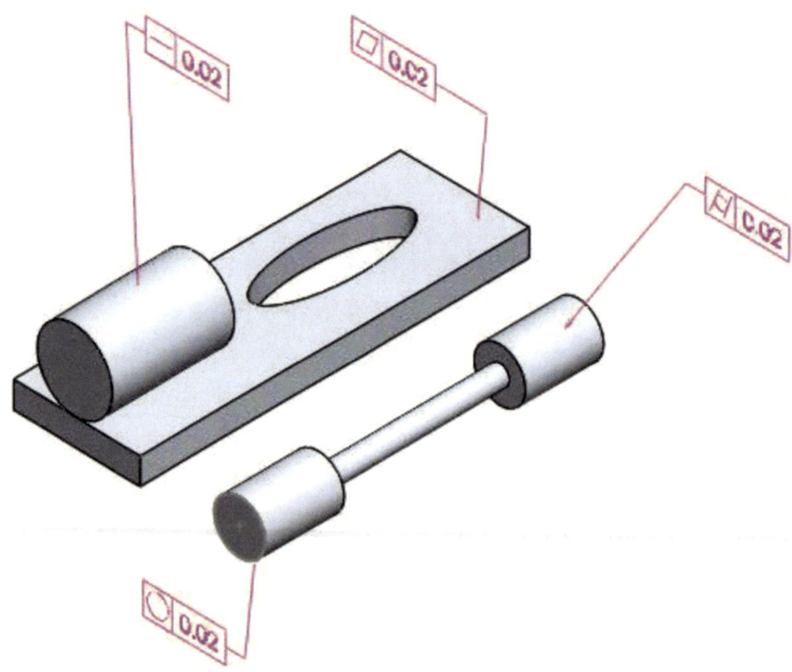

Figure 5.1: Form Tolerances

The Form Tolerances are applied to individual features, element of individual feature of feature of size. So, as it is not related to different feature, Datums are not used with Form Tolerances.

There are four Form tolerances Straightness, Flatness, Circularity and Cylindricity. As shown in figure 5.1. Form tolerances are also controlled by other tolerances such as size tolerance, profile, Runout, or orientation control. So, whenever we use form tolerances consideration must be given to these tolerances and form tolerance value must be smaller than other tolerances.

5.1 Straightness Tolerance:

Straightness is condition of element of surface or derived median line is a straight line. Straightness provides tolerances zone by two parallel lines separated by tolerance value within which element of surface or derived median line must lie.

Straightness must be shown in a view where controlled element shown as line.

Straightness tolerance applied to Surface:

As shown in figure 5.2, the straightness tolerance is applied to the surface or extension of the surface.

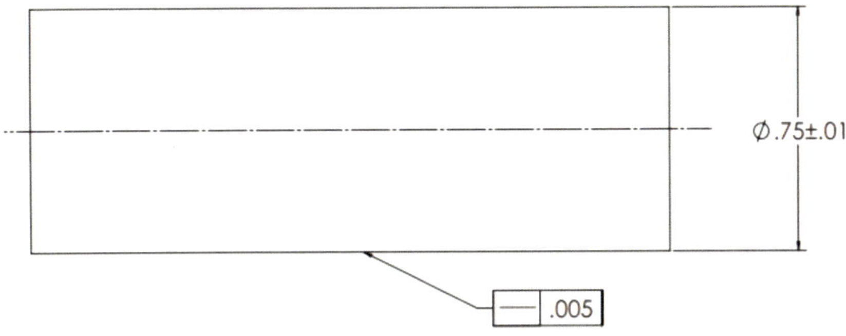

Figure 5.2: Straightness: Applied to the surface

All circular elements must be with in size tolerance diameter .75 ±.01. The longitudinal element must be within two parallel line in a plane common with axis of unrelated mating envelope and separated by tolerance value.

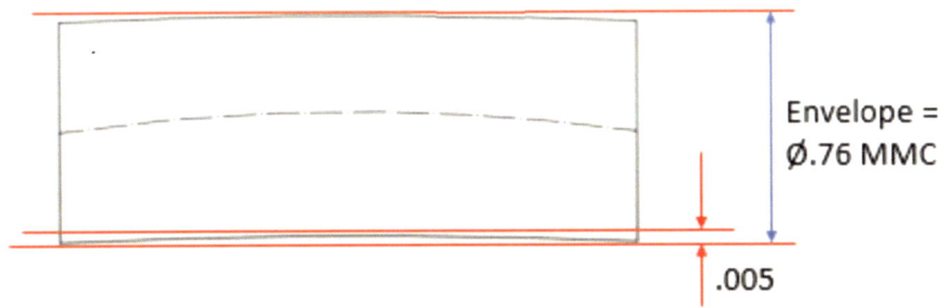

Figure 5.3: Straightness: Tolerance Zone

Straightness Applied to the surface as in Figure 5.2 cannot be on RFS or MMC basis. The tolerance must be less than size tolerance. In this case rule#1 prevails and Size

tolerance controls the straightness. If feature is produced at MMC, perfect form is needed.

In this case:

- ✓ No modifiers.
- ✓ Perfect form at MMC needed.
- ✓ Outer and Inner boundaries remain fixed.
- ✓ Tolerance value must be less then size tolerance.

If Independency symbol applied, Perfect form at MMC not needed. As shown below.

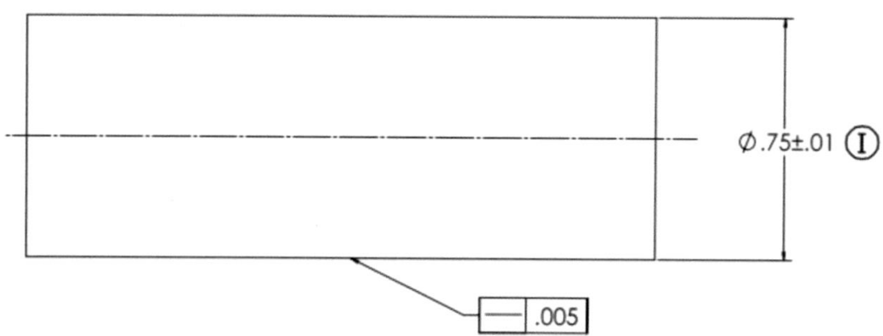

Figure 5.4: Independency Symbol

Straightness tolerance applied on RFS or MMC basis:

When straightness is applied on RFS or MMC basis, feature control frame must be attached to Dimension as shown in figure 5.4.

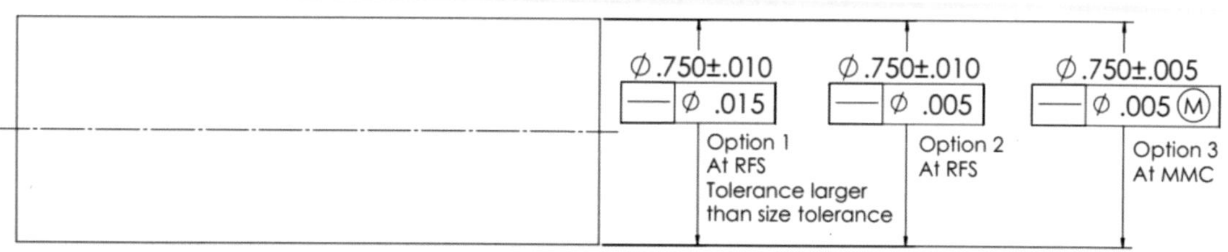

Figure 5.5: Straightness on Feature of size

All the circular elements must follow the size tolerance but perfect form at MMC is not needed. The diameter symbol should be added to feature control frame. If no modifier symbol tolerances are Rule#2 RFS applies.

Figure 5.3 shown Option one has larger tolerance zone; this is only possible if it is not used with any other Geometric Tolerance.

- ✓ No modifiers; so RFS applies.

- ✓ MMC modifier can be used.
- ✓ Boundaries need to be calculated and varies Bonus tolerance introduced.
- ✓ VC = MMC + Geometric Tolerance, RC = LMC – Geometric Tolerance – Bonus (amount of departure)
- ✓ Median line should be derived, and it should be within cylindrical tolerance zone.

If it is RFS, no bonus and tolerance zone remain constant for all produced dimensions. If we use MMC modifier, At MMC size we get only mentioned tolerance but if Feature departs from MMC additional tolerance will be added by the amount of departure as shown in below Table 5.1.

Diameter	Straightness Tolerance
LMC = .740	.025
.745	.02
.750	.015
.758	.007
.759	.006
MMC = .760	.005

Table 5.1: Figure 5.5 Option 3

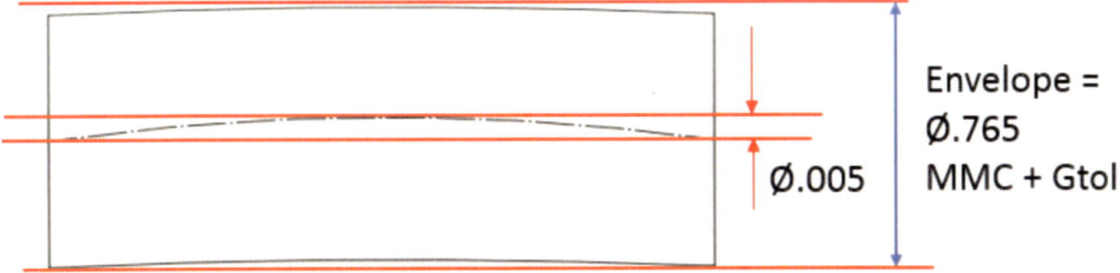

Figure 5.6: Diametric tolerance zone when Feature at MMC

Median line derived from features actual local size should fall within diametrical tolerance zone.

If feature at LMC we get maximum Bonus. The size tolerance must be followed for each local cross section. The Diametric straightness tolerance increases with feature departure from MMC.

Straightness can be applied to Flat surface to control element in particular direction. If we provide tolerance in other view on same surface, both direction elements are controlled. It is as shown if figure 5.7.

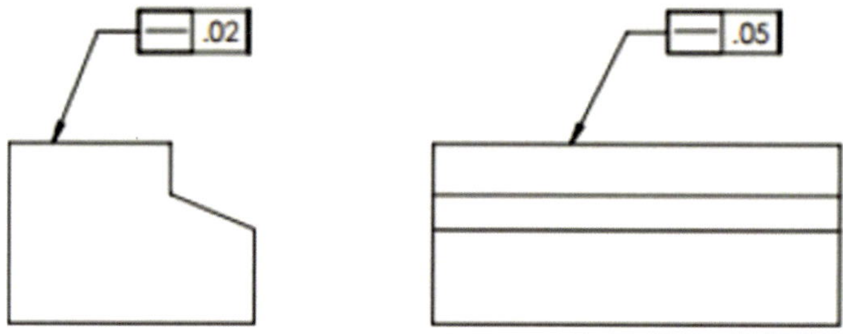

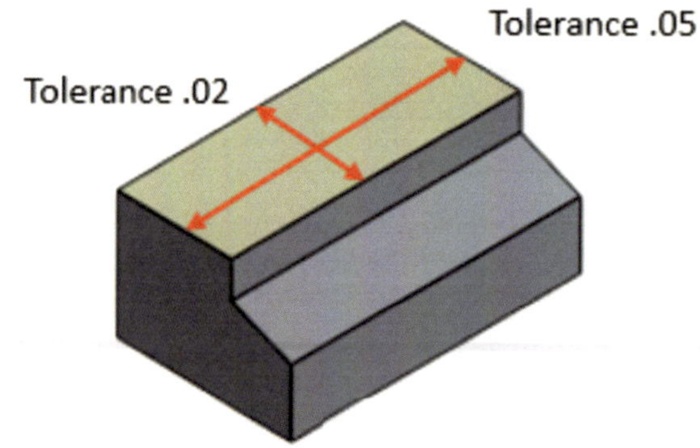

Figure 5.7: On Flat surface

5.2 Flatness Tolerance:

Flatness is Condition of surface or Median plane having all points, elements in one plane.

Flatness tolerance zone is defined by two parallel planes separated by tolerance value within which surface or derived median plane must lie.

When Flatness applied to the surface, the leader is directed to the surface or extension of surface in a drawing view where surface represented as line. In this case rule#1 of perfect form at MMC is required and so the tolerance value must be less than Size

tolerance. Same as straightness tolerance, Independency symbol shall be used where perfect form at MMC not needed and larger flatness tolerance permitted.

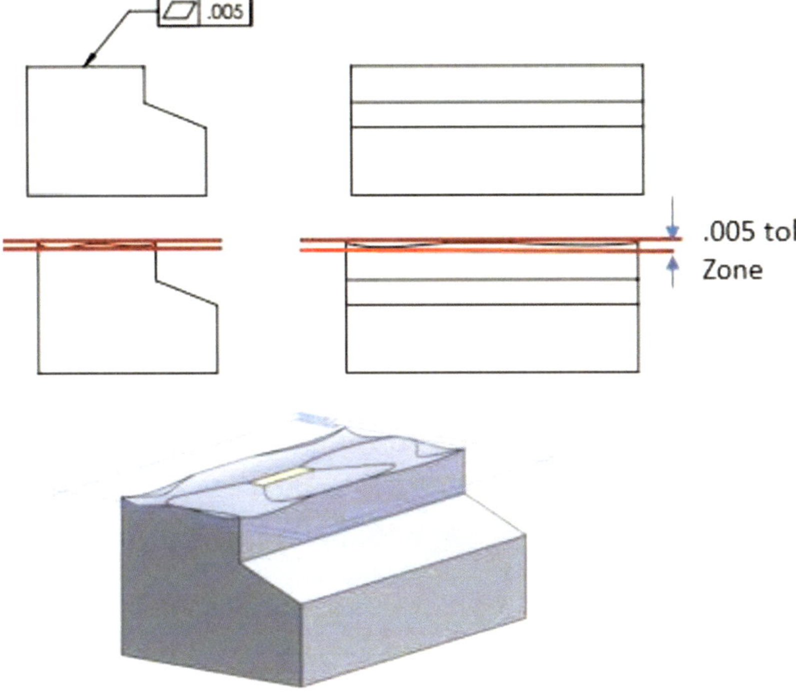

Figure 5.8: Flatness on surface

First plane meets max points on surface, second plane are separated by Geometric Tolerance value .005 and All elements of surface must fall within tolerance zone.

Flatness on RFS, MMC or LMC basis:

Flatness can be applied to feature of size and that makes modifiers RFS MMC or LMC applicable. No diameter symbol is used. The derived median plane must be with in defined tolerance zone.

If it is applied on RFS basis, flatness tolerance for any size of unrelated mating envelope remains constant. If MMC modifier applied in feature control frame, Feature at MMC will have Specified tolerance. If size of unrelated actual mating envelope departs from MMC we get BONUS tolerance.

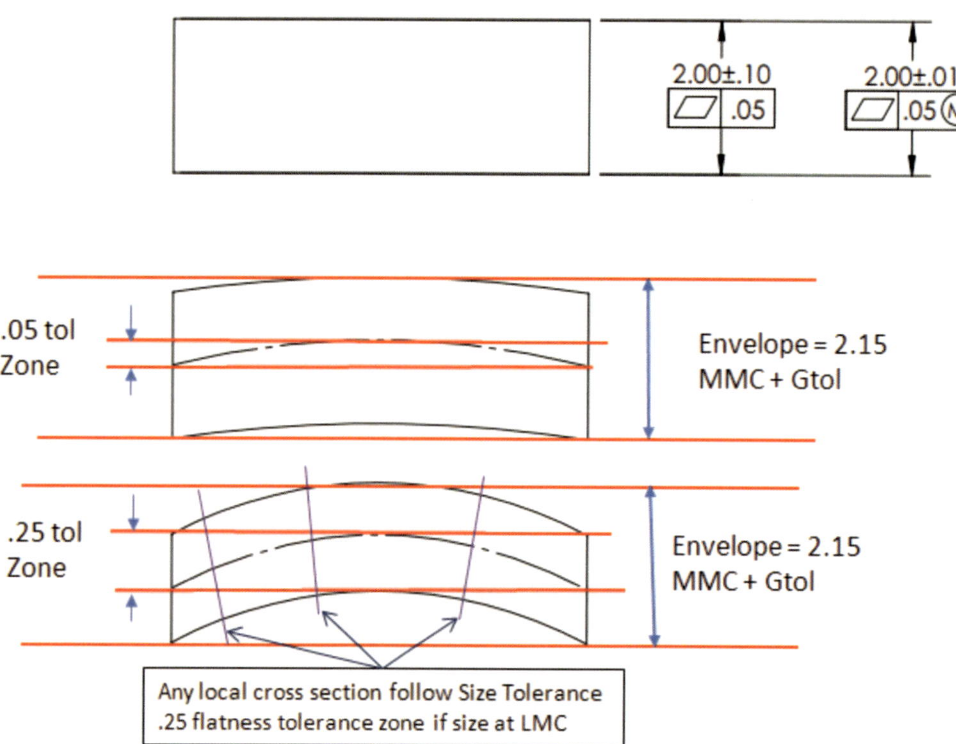

Figure 5.9: Flatness at MMC

Flatness MMC basis tolerance can go up to .25 if manufactured size at LMC. As shown in figure 5.9 and table 5.2.

Diameter	Flatness Tolerance
LMC = 1.9	.25
1.95	.2
2	.15
2.05	.1
MMC = 2.1	.05

Table 5.2: Flatness at MMC

Depending upon requirement, Flatness and Straightness can be applied on per unit basis to avoid abrupt changes on short length or smaller area.

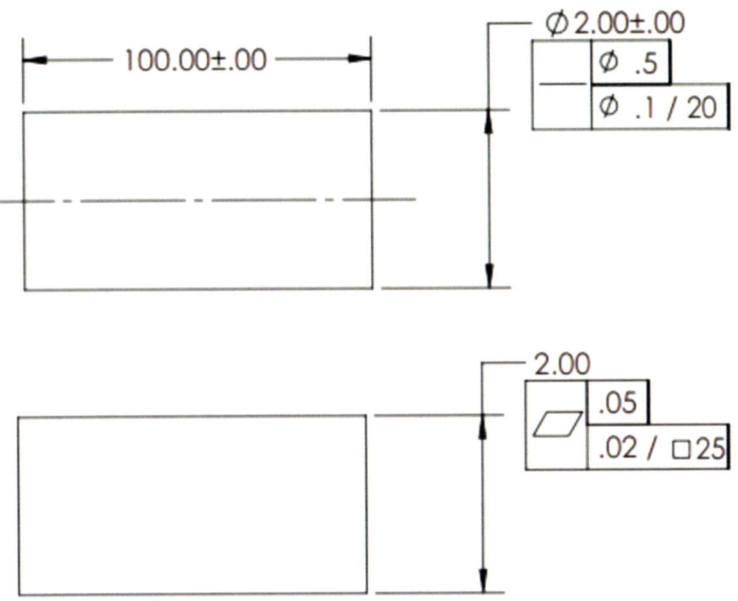

Figure 5.10: Per Unit basis

Composite tolerance, we will discuss in detail with positional tolerancing.

5.3 Circularity Tolerance:

Roundness refers to a condition of a circular line or the surface of a circular feature wherein all points on the line or on the periphery of a plane cross section of the feature, are equidistant from a common center point. For sphere, it is all points intersecting plane passing through center point at equidistance from center point. Figure 5.11 shows some circularity error.

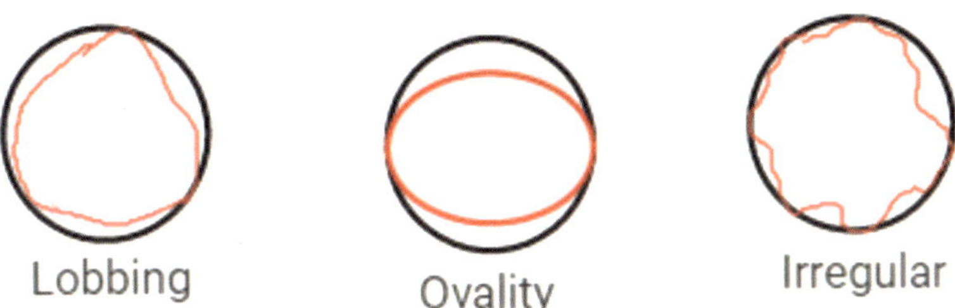

Figure 5.11: Circularity Errors

Circularity tolerance shall be less than size tolerance. The tolerance zone is two concentric circles separated radially by the amount of tolerance. Within this tolerance zone all points on surface at a cross section must lie.

Circularity tolerance may be applied to any type of feature having uniformly circular cross sections, including spheres, cylinders, revolute (such as cones), doughnut shapes, and bent rod and tubular shapes.

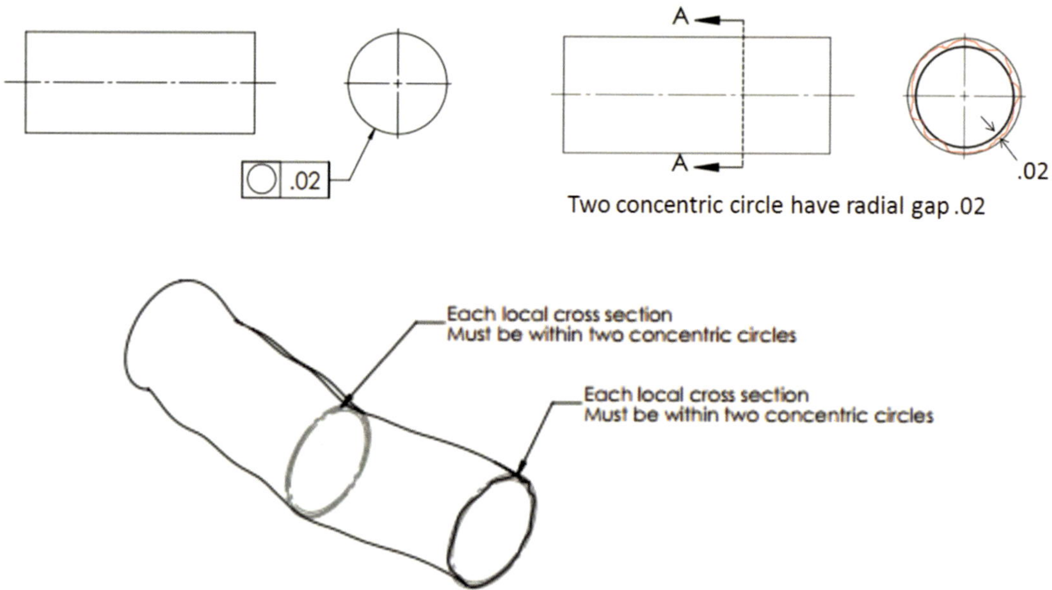

Figure 5.12: Circularity tolerance (for non spherical features)

A circularity tolerance greater than the total size tolerance has no effect as size tolerance to be followed. A circularity tolerance between the full-size tolerance and one-half the size tolerance limits only single-lobed (egg shape) deviations. A circularity tolerance must be less than half the size tolerance to limit multi lobed (such as elliptical and tri-lobed) deviations.

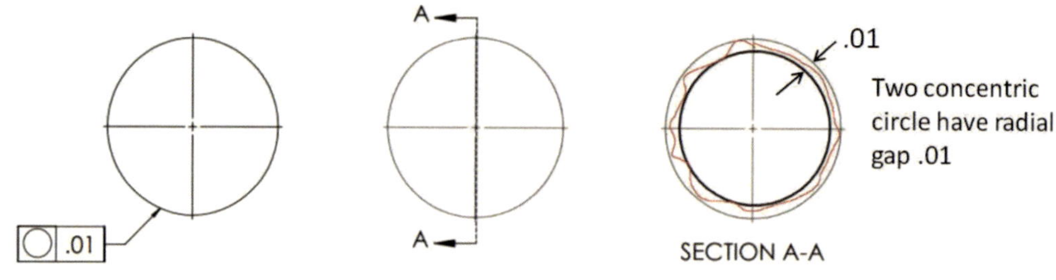

Figure 5.13: Circularity tolerance applied to a spherical feature.

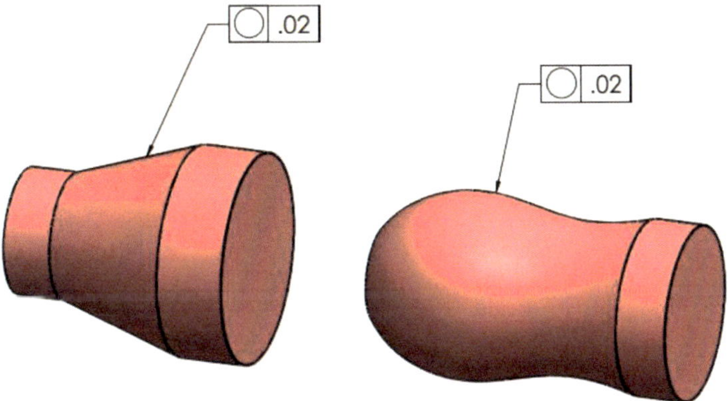

Figure 5.13: Circularity tolerance applied to a non-cylindrical feature.

Circularity tolerance can be applied to the non-cylindrical parts, features as shown in figure 5.13.

Circularity tolerance should be considered in statistical tolerance analysis as it can created slightly larger envelope compared to specified tolerance. General tolerance stack-up analysis does not require considering circularity tolerance as the maximum radial envelop cannot exceed your maximum diameter tolerance due to rule #1 until independency symbol is mentioned.

5.4 Cylindricity Tolerance

Cylindricity tolerances can only be applied to cylindrical surface, such as round holes and shafts. Errors of Cylindricity may be caused by out of roundness, e.g. ovality or lobbing. Errors of straightness caused by bending or by diametric variation or random irregularity.

A Cylindricity tolerance specifies a tolerance zone bounded by two concentric cylinders whose radii differ by an amount equal to the tolerance value. The entire feature surface shall be contained within the tolerance zone (between the two cylinders). The tolerance zone cylinders may adjust to any diameter, provided their radial separation remains equal to the tolerance value.

As with circularity tolerances, a Cylindricity tolerance must be less than half the size tolerance to limit multi-lobed form deviations. Since neither Cylindricity nor circularity tolerance can nullify size limits for a feature. The Cylindricity is applicable on RFS basis and it can be shown in either of views.

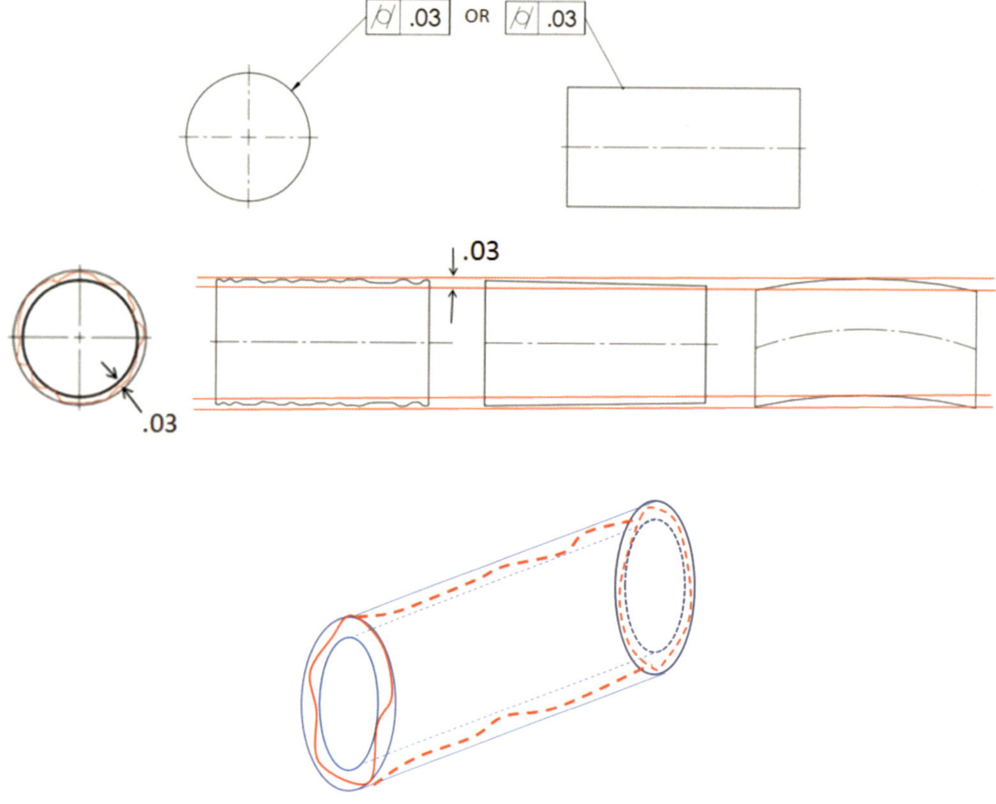

Figure 5.14: Cylindricity tolerance

Cylindricity controls straightness, circularity, and taper of cylindrical feature.

Cylindricity tolerance should be considered in statistical tolerance analysis as it can created slightly larger envelope compared to specified tolerance. General tolerance stack-up analysis does not require considering circularity tolerance as the maximum radial envelop cannot exceed your maximum diameter tolerance due to rule #1 until independency symbol is mentioned.

5.5 Application on Limited Length or Area

Some designs require form control over a limited length or area of the surface, rather than the entire surface. In such cases, draw a heavy chain line adjacent to the surface, basically dimensioned for length and location, as necessary. See Figure 5.15.. The form tolerance applies only within the limits indicated by the chain line.

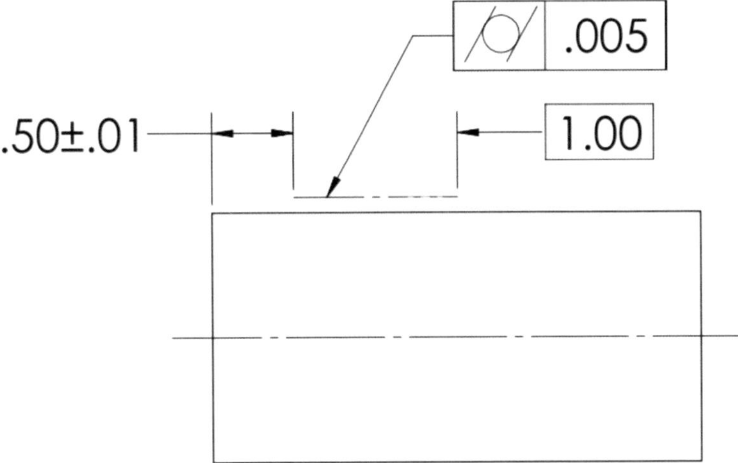

Figure 5.15: Cylindricity tolerance applied over a limited length.

Chapter 6

Datum Selection:

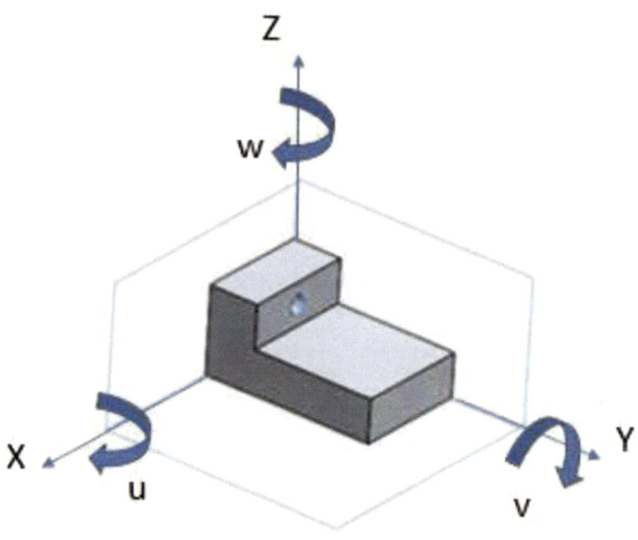

Figure 6.1: Degree of freedom

The Purpose of datum is to create origin for Relative Dimensions with reference to which Rotational and Translation degree of freedom is locked. We will discuss method of establishing datum feature and datum reference frame.

Datum can be axis, point, line, or Plane. These datums or combination are used in order of primary, secondary, and tertiary datum and establishes datum reference frame (three mutually perpendicular frames).

6.1 Datum Feature:

The surface or feature of size provides origin for measurement and plies crucial role in assembly is selected as Datum Feature. Only required datum feature should be mentioned in feature control frame. Proper understanding of geometric tolerancing is required while selecting Datum feature.

A datum feature is selected on basis of design requirements. The mating surfaces are generally selected as Datum feature. Sufficient length is needed to simulate Datum reference from datum feature. Datum feature must be easily located on drawing. If in case of twin identical feature, need of marking feature may arrive for identification.

Geometric tolerance does not consider any form or location variation of datum features. Datum features shall be controlled by Geometric Tolerances directly or it should be controlled by size tolerance. Primary Datum feature's form and location of pattern used establish Primary Datum. Secondary or Tertiary Orientation Datum features orientation or location to higher precedence datum.

6.2 Datum Identification:

Datum features are selected based on functional assembly and mating conditions.

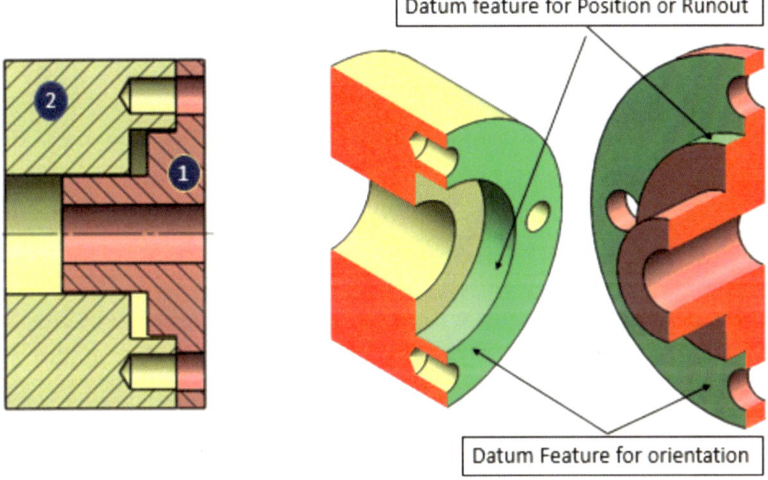

Figure 6.2: Assembly

Figure 6.2 shows component 1 and 2 are assembled by bolts (not shown in assembly) The flat face of the part 1 touches with part 2 and orientates the parts relatively so face can be selected as datum feature as shown in figure 6.3. Highlighted feature (in green)

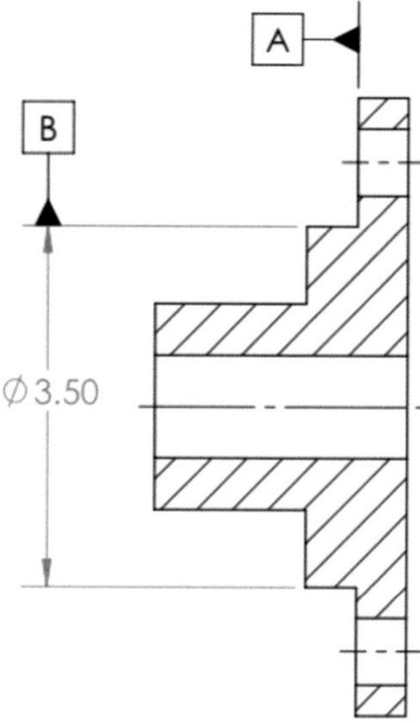

Figure 6.3: Functional Datums

(boss) locates the feature and in contact with other part bore is selected as the datum B. As the assembly rotation is locked by bolts, no tertiary datum needed in this example.

6.3 Temporary Datum and Permanent Datums:

The Feature those in process and will be machine later e.g. casting face or forged feature, these can be selected as temporary datums and will be machined in subsequent machining processes. The Permanent datum feature should not be changed in any subsequent machining processes.

6.4 Datum Reference Frame:

Mutually perpendicular planes set as reference frame are simulated from datum features to provide origin of measurement. This constrains translation and rotational degree of freedom. Depending on requirement there may be single or multiple datum reference. As shown in figure 6.4. Different datum of precedence or material boundaries create different Datum reference frame.

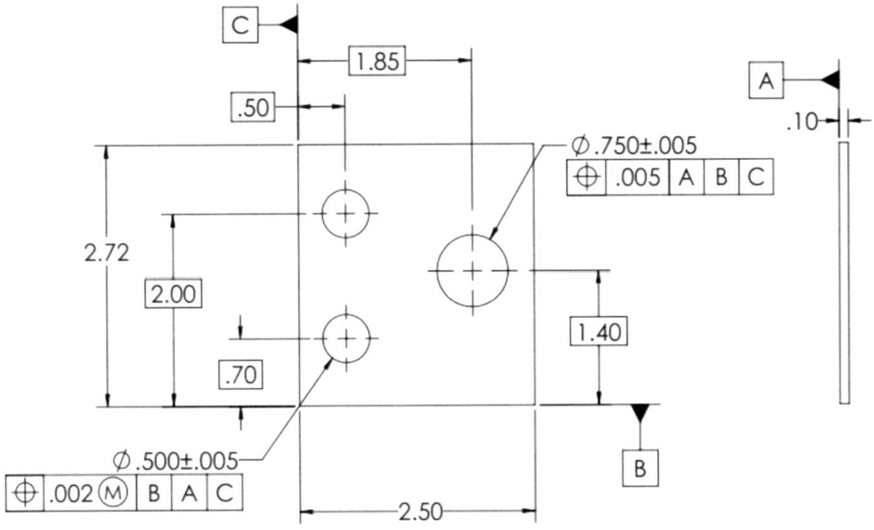

Figure 6.4: Multiple Datum reference Frame

We will discuss this furthermore in simultaneous requirement.

6.5 Datum Simulator:

The inspection equipment used to establish datums are called as physical datum feature simulator. Datum feature simulators discussed in ASME 2009 Y14.5 are theoretical datum feature simulator. Datum simulator is inverse shape of the datum feature used to established datums and datum reference frame. Depending on datum feature, the simulator may be:

MMB, LMB

Actual mating envelope

Minimum material envelope

Mathematical boundary

Datum Target,

And Tangent plane.

The Datum simulators should have perfect form, basic location with reference to other simulators, Basic orientation within all simulators, Movement allowed if translation symbol, fixed in size if RMB or LMB and Adjustable in size if at RMB.

Simulated datum is Plane, Axis or Point established by datum feature simulator.

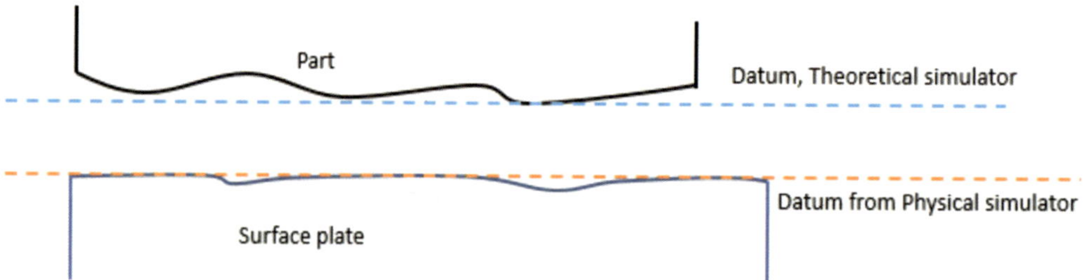

Figure 6.5: Datum Simulator

6.6 Planar Datums:

A Planar Datums are derived from planar datum feature, generally planar surfaces.

Planar datum features are selected function and design requirement. Figure 6.3 shows Datum A is selected based on interfacing surface in assembly. Mainly Planar Datums orientates and locates the part or feature in an assembly.

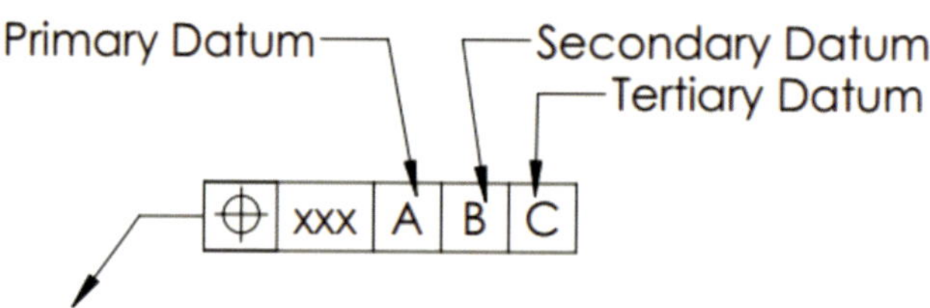

Figure 6.6: Feature control frame

Primary Datum – First datum feature to contact the gage during inspection, Secondary Datum – Second datum feature to contact the gage and Tertiary Datum – Third datum feature to contact gage and hence order of precedence followed and degree of freedom constrained accordingly.

When Datum feature is flat surface, the datum simulator is a plane. Datum feature simulator contacting number of point depends whether datum is primary, secondary or tertiary see figure 6.7.

Rule 3-2-1 applied to these datums in order of precedence.

The Primary datum established by datum feature simulator makes maximum contact with datum feature and minimum at three points. This constrains 3 degree of freedom, one translation and 2 rotations.

The Secondary datum established by simulator makes minimum two points contact and restricts one rotation and one translation.

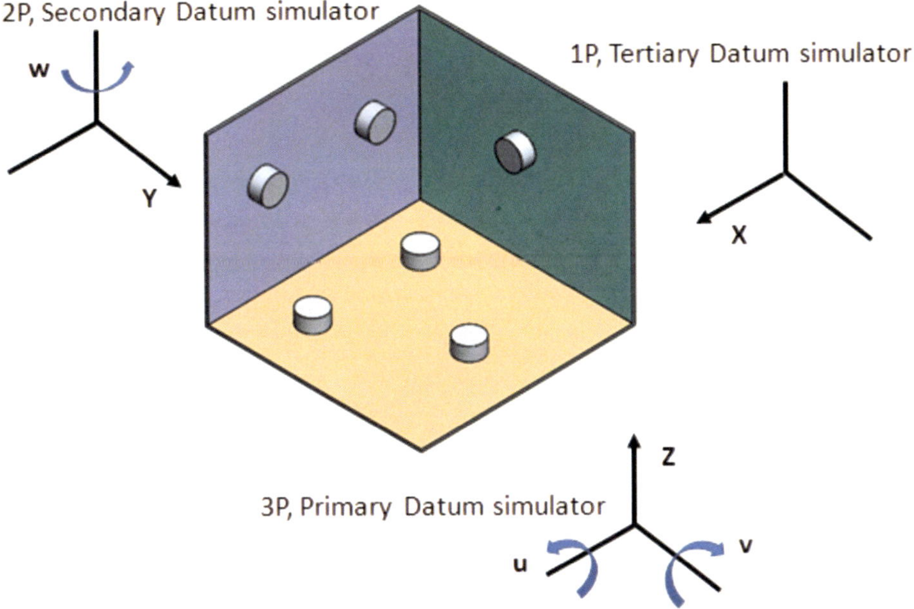

Figure 6.7: 3-2-1 Principle

The Tertiary Datum established by simulator makes minimum one points contact and restricts one translation.

This is how by mobilizing datum reference frame, this restricts all 6 degree of freedom and creates origin for measurement.

Example is tripod stands; three-point contact needed to stabilize it.

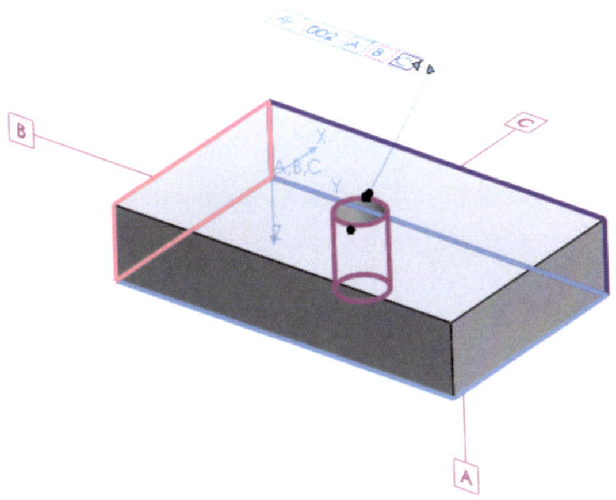

Figure 6.8: Development of Reference frame

Figure 6.7 shows the Geometric Tolerance with reference to the planar datum features. Whenever surface is made as primary datum feature high points of the surface establishes the Datum. Minimum three points of feature come in contact with simulator and establishes datum. This restricts three degree of freedom. One translation and two rotations.

Secondary Datum feature B contacts at minimum two points to simulator and establishes datum. This restricts 2 degree of freedom. One translation and one rotation.

Tertiary Datum feature C contacts at minimum one point and restrict one translation degree of freedom. This establishes datum reference frame.

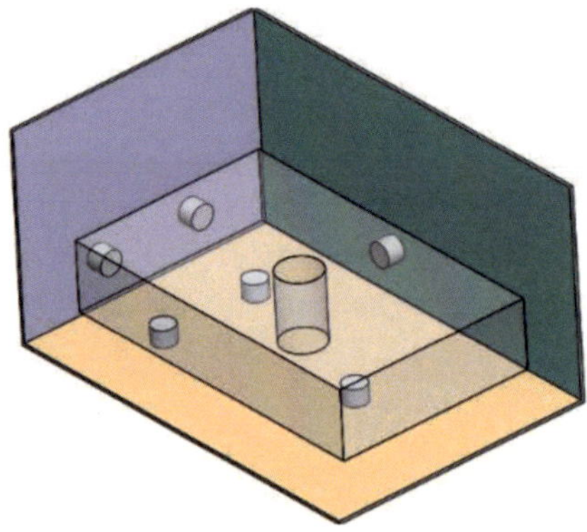

Inclined surface:

In the case of inclined surface, simulator orientated at basic angle as defied in drawing. The Figure 6.9: Simulator and Reference frame

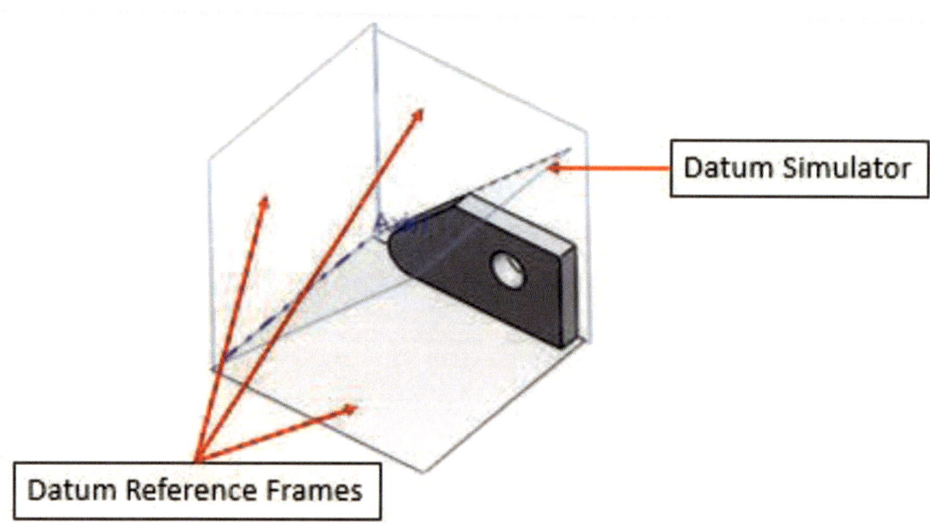

Datum is plane of reference frame passes thru vertex of angle and perpendicular to the other datum references.

6.7 Datum feature at RMB:

Whenever Datum feature is at RMB, the simulator originates at MMB and expands or contracts toward LMB and makes maximum contact with surface. Practical example is chuck, centering devices, V blocks etc.

Diameter:

If diameter is primary datum feature, the axis of datum feature simulator is the Datum. The datum feature simulator is unrelated actual mating envelop, smallest contacting diameter for external feature and largest contacting diameter for internal feature making maximum contact with datum feature.

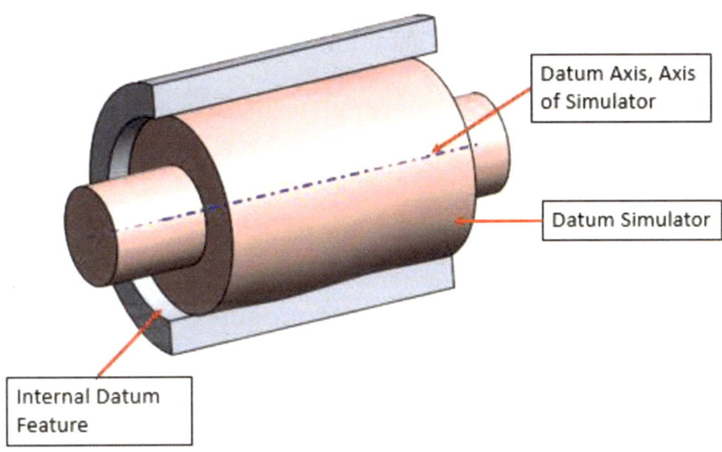

Figure 6.10: Internal Datum feature and Simulator

This axis sets origin for measurement and reference for Geometric Tolerances. This establishes two reference datum passes through the axis. It will not be constrained in rotation will need another datum to constrain rotation degree of freedom.

If the diameter or cylindrical feature is secondary or tertiary datum, it is established in same way as primary datum but its orientation with Datum of precedence must be followed.

Width:

If datum feature is width, the center plane of the feature simulator is datum. The simulator is two planes with minimum separation for external feature and maximum separation for internal feature.

This expands or contracts and makes maximum contact with the surface and establish datum center plane.

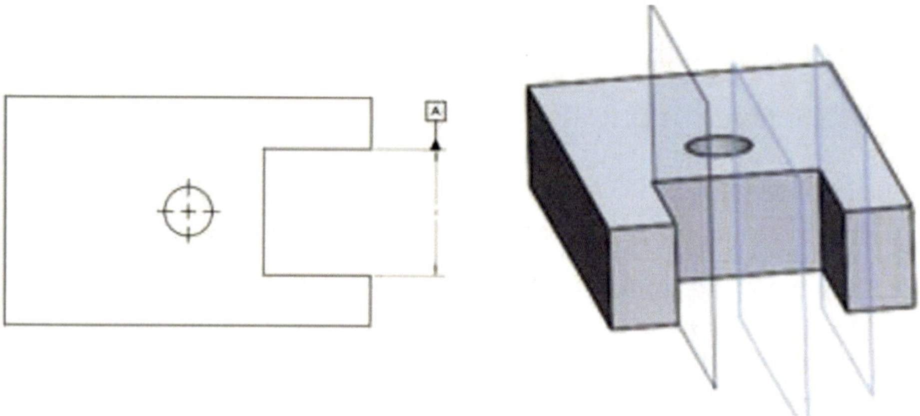

Figure 6.11: Width as Datum Feature

If width datum is secondary or tertiary, simulator is basically oriented to the datum of precedence.

Surface as Secondary or Tertiary Datum feature at RMB:

Surface as secondary or tertiary datum feature simulator originates at MMB and moves towards LMB and makes maximum contact with surface. We will see these examples in section of constraining rotational degree of freedom.

Spherical feature can be primary or secondary datum feature based on functional requirement.

6.8 Maximum Material boundary (MMB):

The maximum material boundary can be applied to datum by adding MMC modifier to the datum symbol in feature control frame as shown in figure 6.11. In this case datum simulator size is fixed and can be determined by mmc condition of datum feature and effect of any datum of precedence.

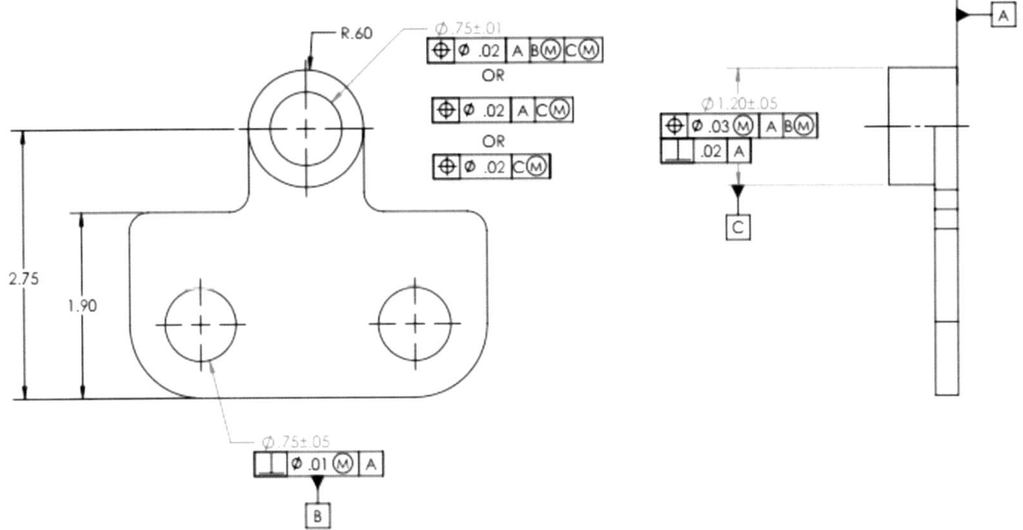

Figure 6.12: Datum Feature at MMB

For Internal feature, it is largest MMB and for external feature it is smallest MMB considers precedence relation. Figure 6.12 shows MMB with datum. Let us find out different envelope size for simulator for Datum C with different applications:

$$\boxed{\oplus \;\middle|\; \varnothing\; .02 \;\middle|\; A \;\middle|\; B\text{\textcircled{M}} \;\middle|\; C\text{\textcircled{M}}}$$

Find datum feature simulator size when A and B datums are primary and secondary datum.

MMC for Datum C = 1.25

Effect of Datum of precedence = Geometric Tolerance on the datum feature related to these datums = .03

Simulator Size = 1.25+.03 = 1.28

$$\boxed{\oplus \;\middle|\; \varnothing\; .02 \;\middle|\; A \;\middle|\; C\text{\textcircled{M}}}$$

Find datum feature simulator size when A datum is primary datum.

MMC for Datum C = 1.25

Effect of Datum of precedence = Geometric Tolerance on the datum feature related to these datums = .02

Simulator Size = 1.25+.02 = 1.27

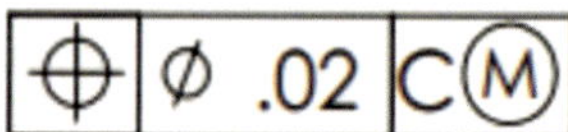

Find datum feature simulator size when C datum is primary datum.

MMC for Datum C = 1.25

Simulator Size = 1.25 (no other datum of precedence).

Similarly, it can be calculated for other datums. If we want to restrict these sizes or make it basic, we can do that with a not in feature control frame.

6.9 Least Material boundary (LMB):

The least material boundary can be applied to datum by adding LMC modifier to the datum symbol in feature control frame. In this case datum simulator size is fixed and can be determined by LMC condition of datum feature and effect of any datum of precedence.

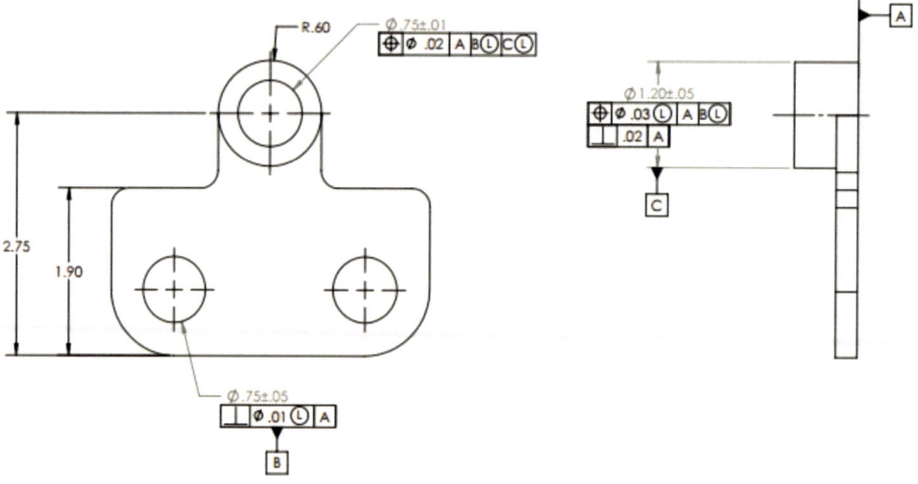

Figure 6.13: Datum Feature at LMB

Find datum feature simulator size when A and B datums are primary and secondary datum.

LMC for Datum C = 1.15

Effect of Datum of precedence = Geometric Tolerance on the datum feature related to these datums = .03

Simulator Size = 1.15-.03 = 1.12

6.10 Datum Shift:

Figure 6.14 show that simulator size fixed and for datum C it is 1.28. If the datum feature manufactured at LMC size 1.15 we get datum shift of 1.28-1.15 = .13 though this do not affect Geometric Tolerance anyway, but this should be considered in stack-up analysis.

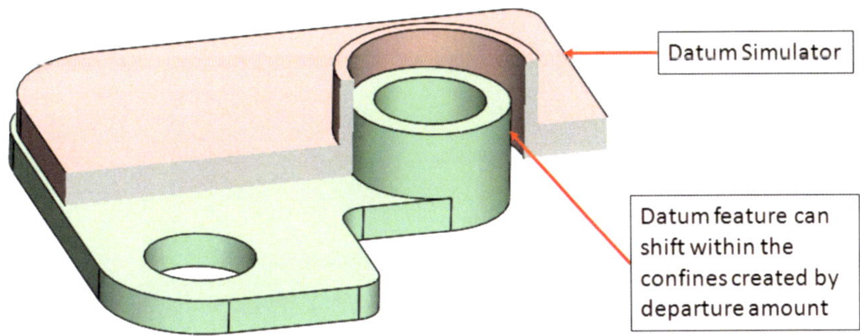

Figure 6.14: Datum Shift

6.11 Translation Modifier:

Translation modifier unlocks basic location and datum feature simulator allowed to translate with in Geometric Tolerance to fully engage the feature. This allows establishing good contact.

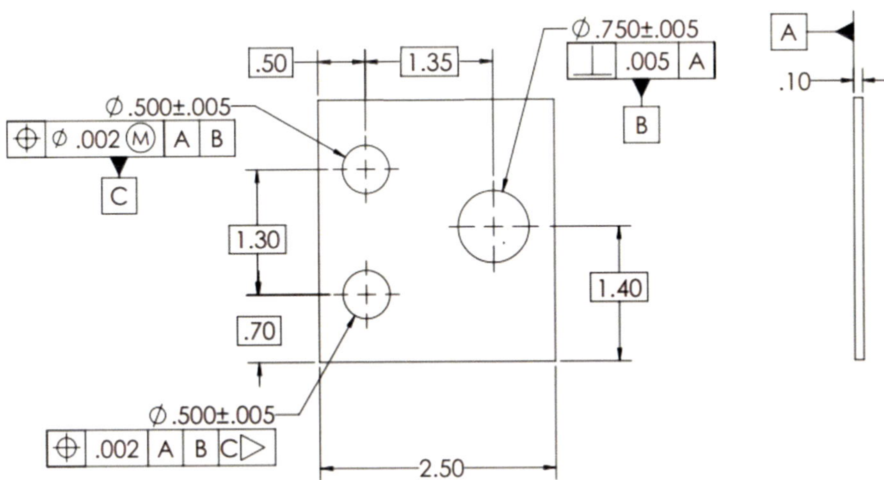

Figure 6.15: Datum translation modifier

This allows Datum C can move w.r.t. datum B. Movement direction may be needed in some cases.

6.12 Effect of shuffling datums in Feature control Frame:

Considering Following example figure **6.16** and let us discuss effect of shuffling datums:

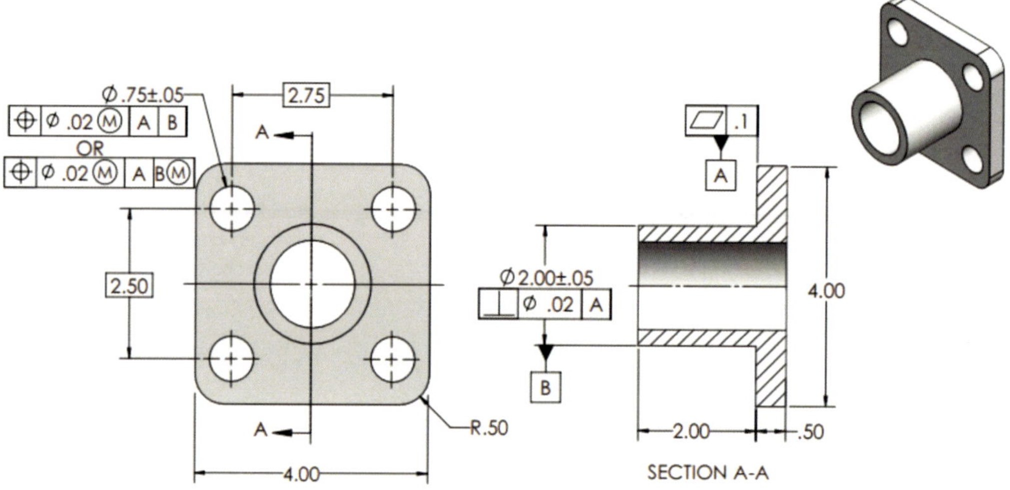

Figure 6.16: Surface as Primary Datum

Figure 6.16 shows surface as primary datum. Once the surface is established i.e., datum A, Datum simulator for B (related actual mating envelope) perpendicular to the Datum A establishes datum axis B.

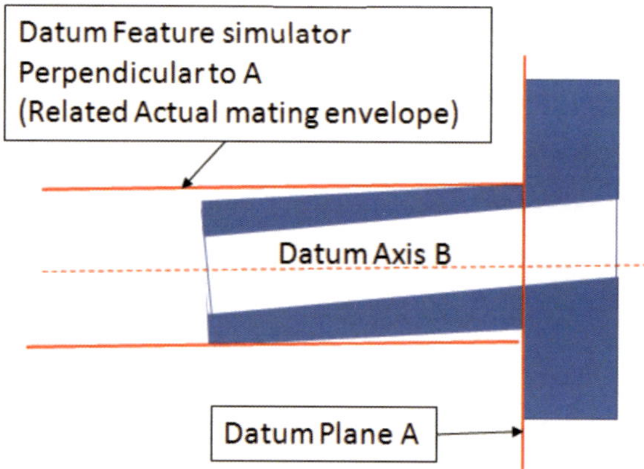

Figure 6.17: Datum Simulators.

If same case is with MMB datum B:

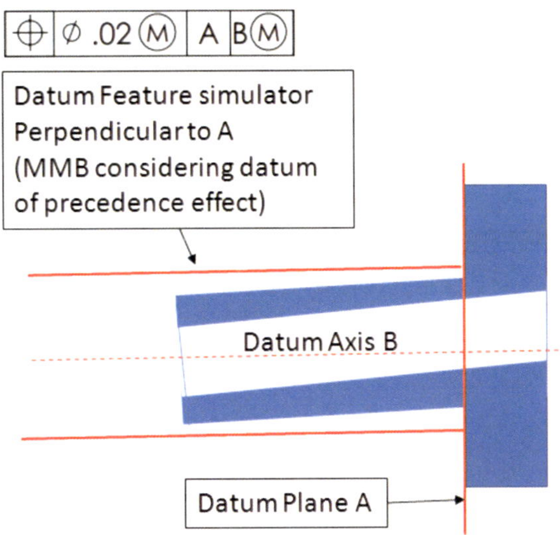

Figure 6.18: Datum Simulators.

Datum Feature simulator diameter is fixed = MMC + Geometric Tolerance of datum of precedence = 2.05+.02 = 2.07.

Let see cylindrical feature as Primary Datum:

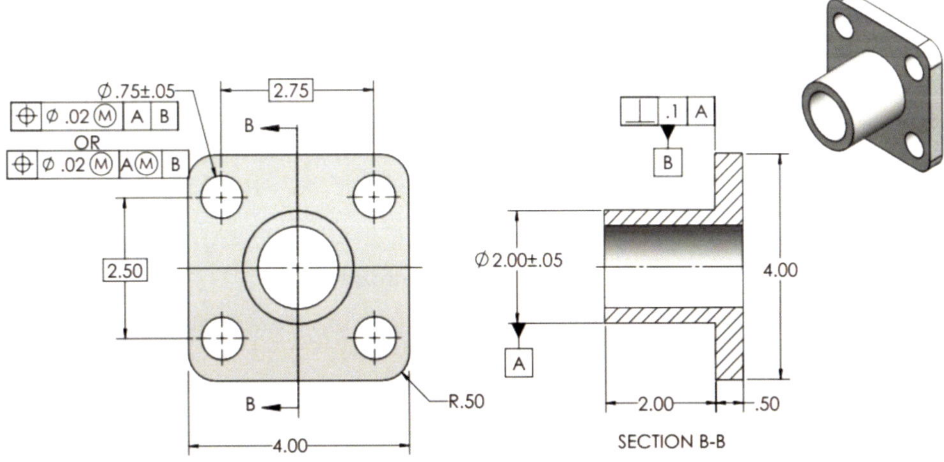

Figure 6.19: Datum Simulators.

Figure 6.19 shows cylindrical feature as primary datum. The Unrelated actual mating envelope is datum feature simulator establishes primary Datum axis A, the Datum Feature simulator plane is now perpendicular to Datum A makes maximum contact

with surface to establish datum B, this affects points of contact as shown in Figure 6.20.

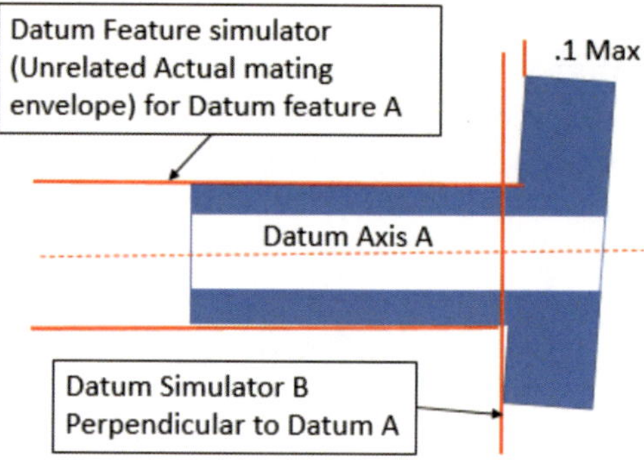

Figure 6.20: Datum Simulators.

If Datum is at MMB:

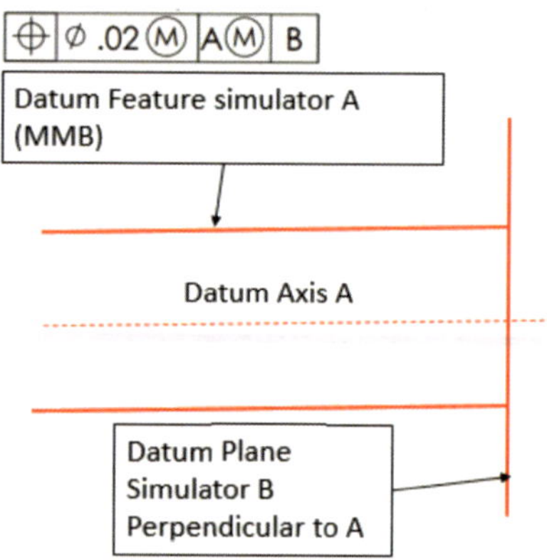

Figure 6.21: Datum Simulators.

Datum Feature simulator diameter is fixed = MMC = 2.05

6.13 Datum Targets:

In case entire feature or surface cannot be used as datum due to irregularities in surface, we use datum targets. Datum Targets are points, Line, or area. For example casting surfaces, non-planar uneven surfaces. Datum Targets are dimensioned and located basically on drawing as shown in figure (6.22).

A primary Datum is established by three datum targets, secondary established by two datum targets and tertiary by one target. If it is irregular or stepped surface must have at least one target.

Depending on requirement moveable datum target may be used to show movement of datum simulator. If the simulator is at RMB, movement of simulator is there but need to represent by a symbol. But it can be given for clarity. If movement is not normal to profile, it needs to be mentioned as shown in **figure 6.22**

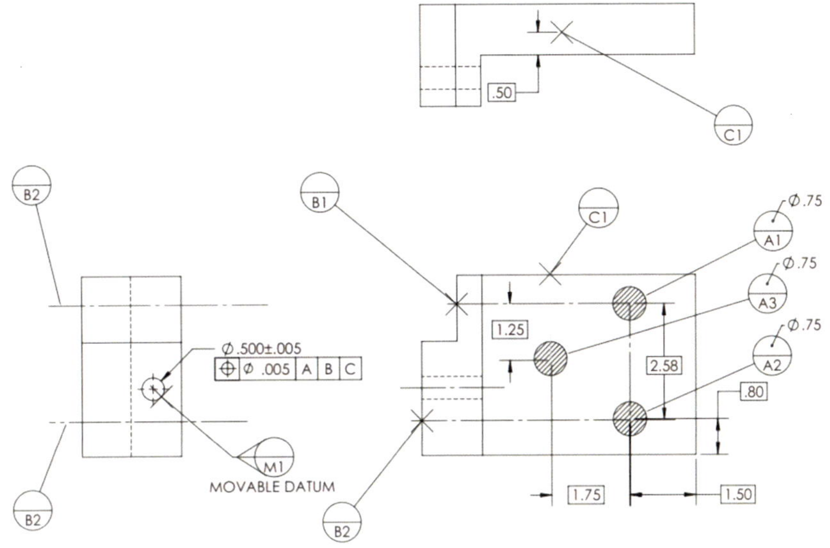

Figure 6.22: Datum Targets – Movable datum symbol for reference only

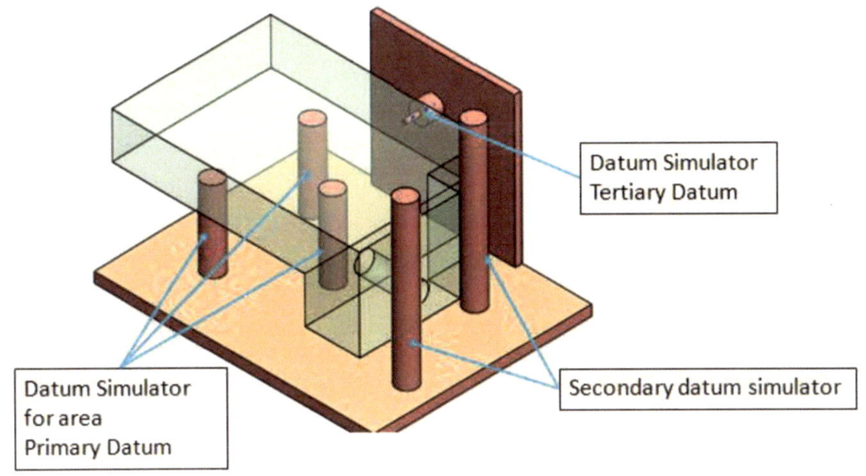

Figure 6.23: Datum Targets.

6.14 Single axis of two coaxial features Or Single Datum plane:

Two stepped surfaced of different diameters are used to established single datum, the datum letters are separated by dash – in single feature frame as shown in figure 6.24. Both datum feature simulator establishes single datum plane or axis.

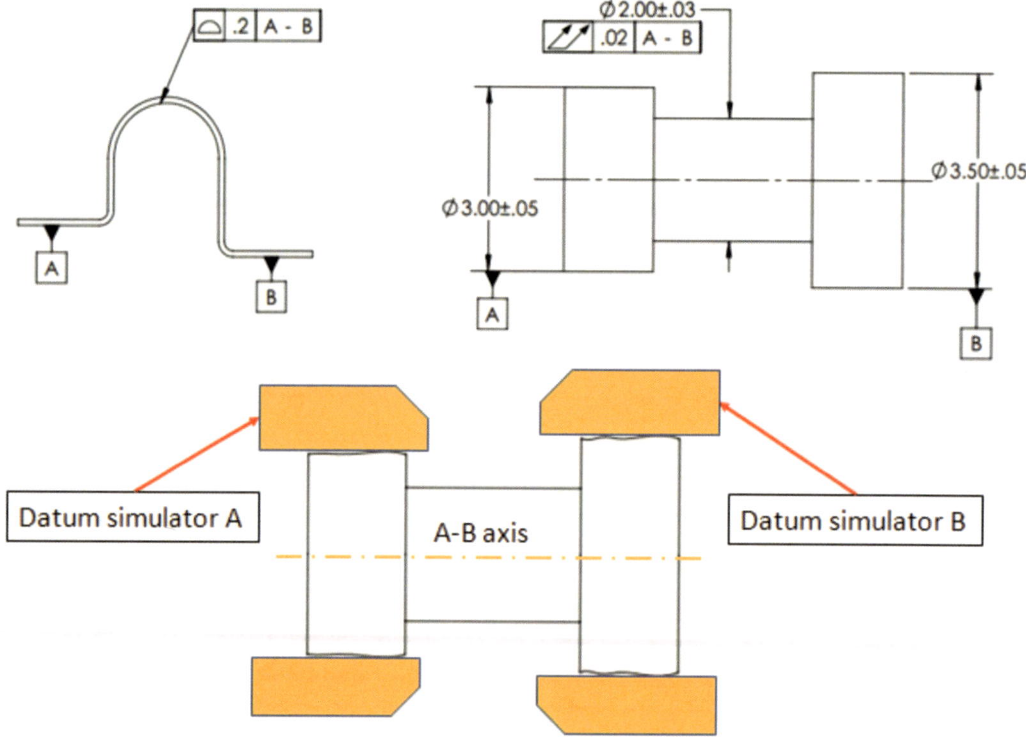

Figure 6.24: Co-Axial Datum

If the surface of diameter is not continuous and both surfaces needed to be considered for datum, it can be defined as shown figure.

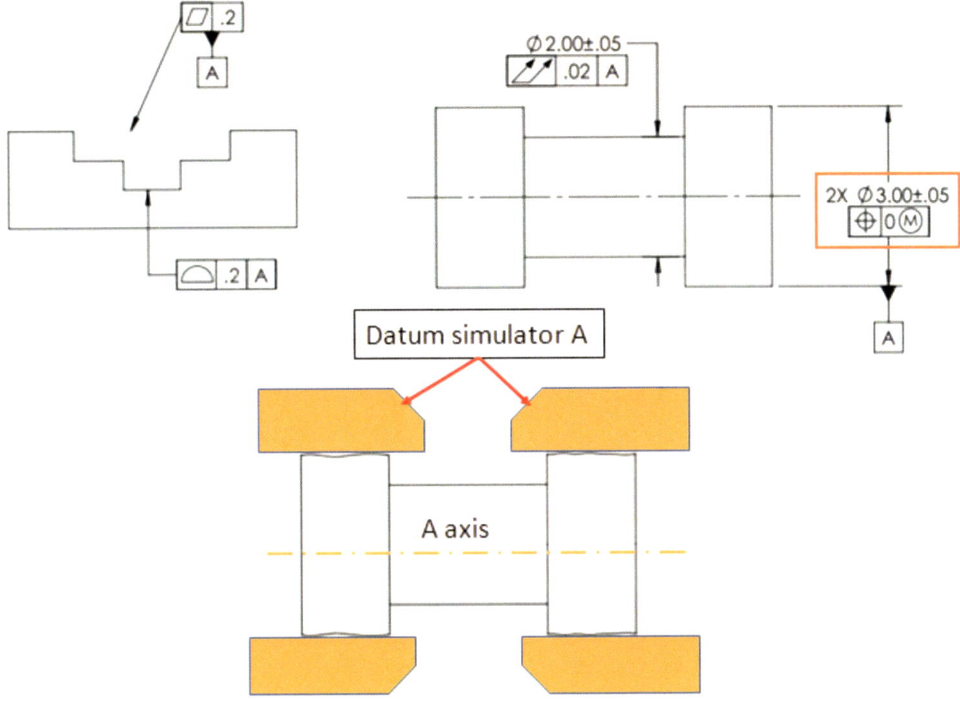

Figure 6.25: Co-Axial Datum

The highlighted Geometric Tolerance for co-axial feature need not to have Geometric Tolerance. We will discuss in detail in Chapter of location tolerance.

6.15 Partial surface of Datum:

We can use partial surface to establish a datum as shown in figure:

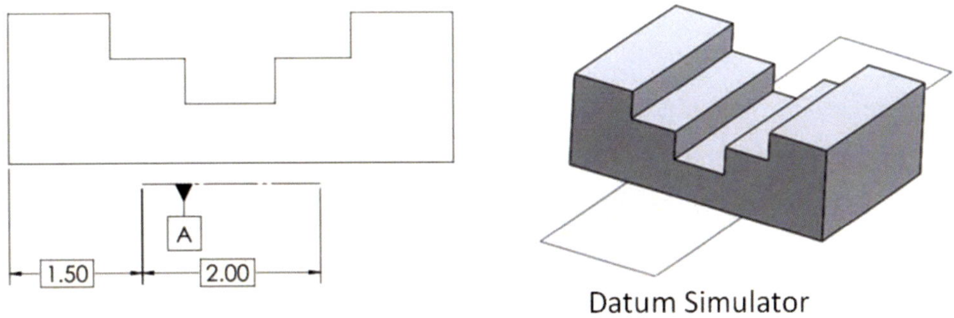

Figure 6.26: Partial surface Datum

If the surface is large, it is desirable to use partial surface to establish datum. It can be represented on drawing as shown in figure 6.26

6.16 Rotational constrain due to axis:

Where datum is a point or axis established by Primary or secondary datum, the datum do not has any rotational constrain. To constrain rotation degree of freedom, lower precedence of datum is used.

Depending on functional requirement, there may be way to constrain rotational degree of freedom as shown in figure 6.27.

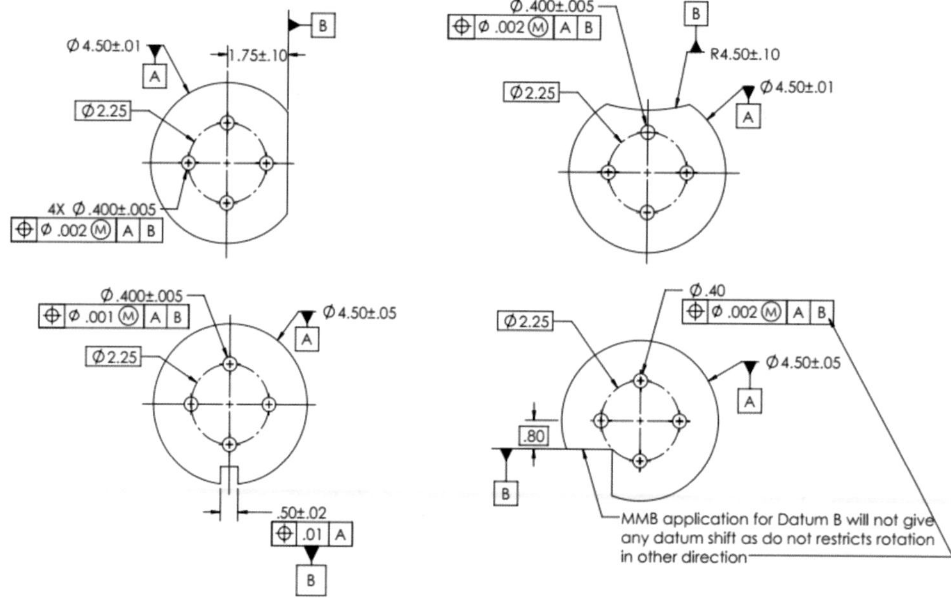

Figure 6.27: Rotational constrain

As shown, Flat surface, a curved surface or a feature of size can be used to constrain rotational degree of freedom.

If these are at RMB, the simulator originates at MMB and moves towards LMB to make maximum contact to established datum.

If at MMB, simulator is fixed at MMB and datum shift may be available depending upon departure amount, but it still locks the rotation.

If datum B does not constrain Rotation in both directions, even if MMB applied movement will not be permitted (no datum shift). Simulator has to be in contact with datum feature.

6.17 Simultaneous requirement:

Where two or more Geometric Tolerances applied as single pattern or part requirement, simultaneous Requirements comes in picture.

Simultaneous requirement applies to Location and profile tolerances. It needs:

- Basic dimension for location
- Common datum feature in feature control frames
- Same order of precedence
- Same material boundary

In this Geometric Tolerances applied to features considered as related groups, no translation or rotation between references frames are permitted.

If same interrelationship not needed, SEP REQ requirement is mentioned. As shown in figure 6.28 and 6.29.

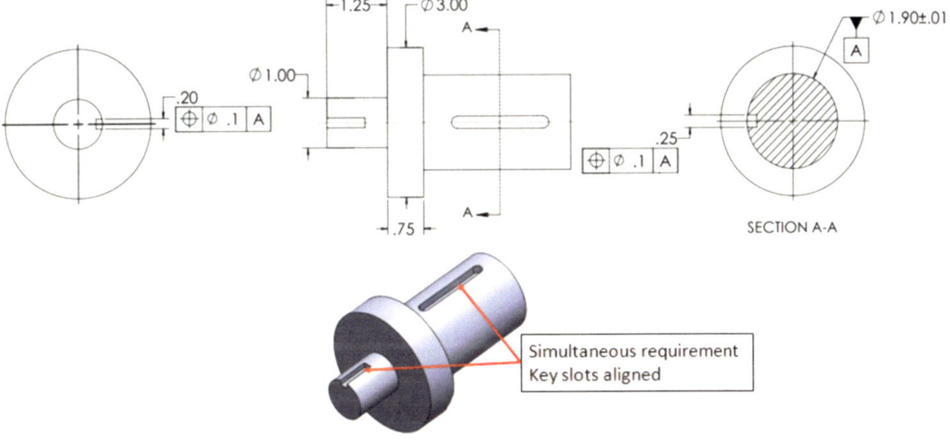

Figure 6.28: Simultaneous Requirement

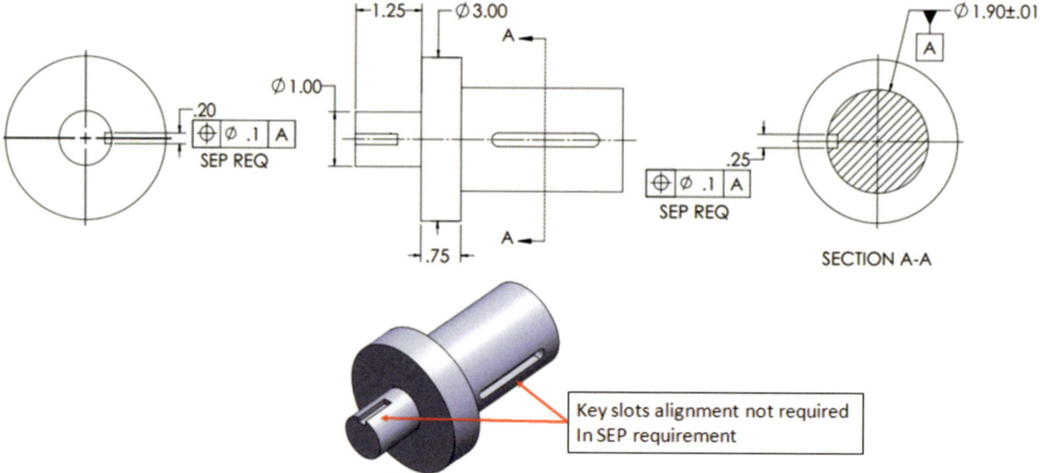

Figure 6.29: SEP REQ, Separate Requirement

This principle is not applied to lower segment of composite tolerances.

6.18 Restrained condition

It is always Free State unless mentioned. It may be required restrain part on its datum to simulate functional or interrelation requirement. Note must be given on drawing for specific requirement. When part is restrained during manufacturing and Forces of manufacturing release part shape or size changes. During inspection the actual assembly condition or restrained condition may be used.

6.19 Identifications:

It may be required to clarify in a drawing view for axes of reference frame, center planes labeled as X, Y and Z (Translation degree of freedom) and (ABC) (ABD) etc. datum mentioned for different frames. Modifiers not needed to be mentioned.

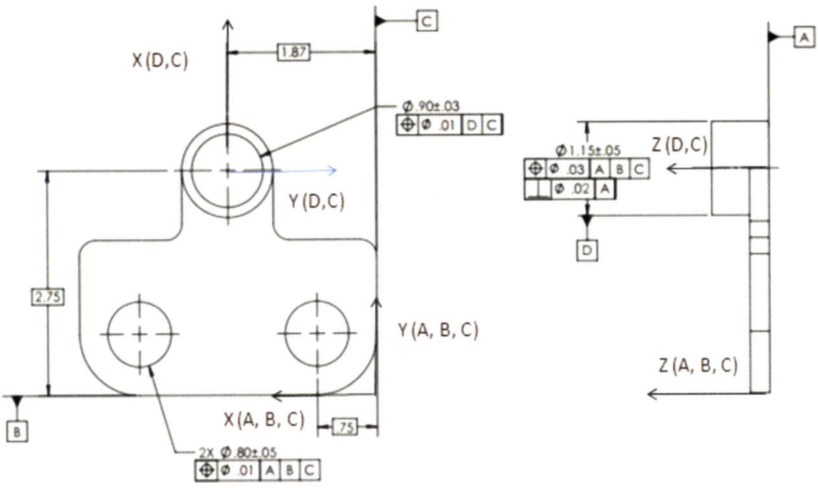

Figure 6.30: Datum Identification

6.20 Customized reference frame:

When it is required to override degree of freedom constrained by a datum, customized reference frame is used as shown in figure 6.31

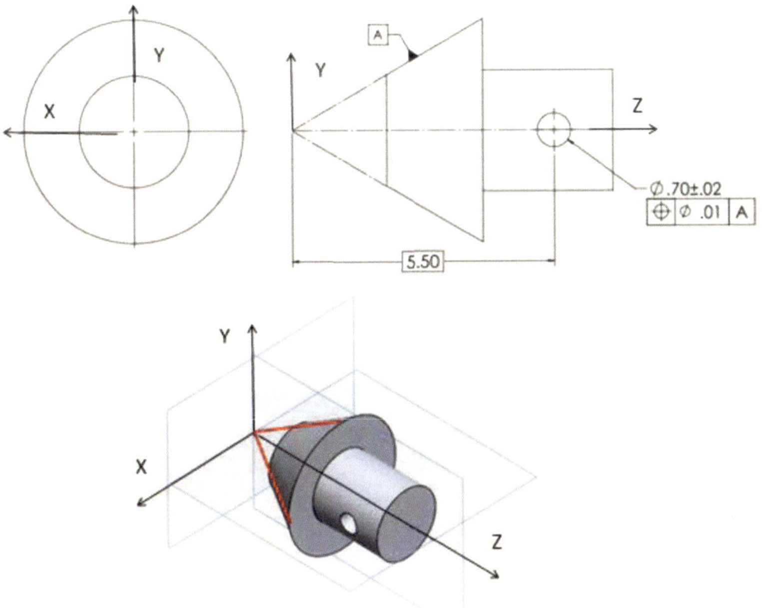

Figure 6.31: Conical Datum Feature

The Conical datum feature established a datum feature simulator results in a point and axis and hence constraining five degree of freedom All translation and two rotation.

Now if we want to have Translation in Z to be restricted in other datum surface as shown in figure 6.32, the customized feature control frame must be established for clarification as shown.

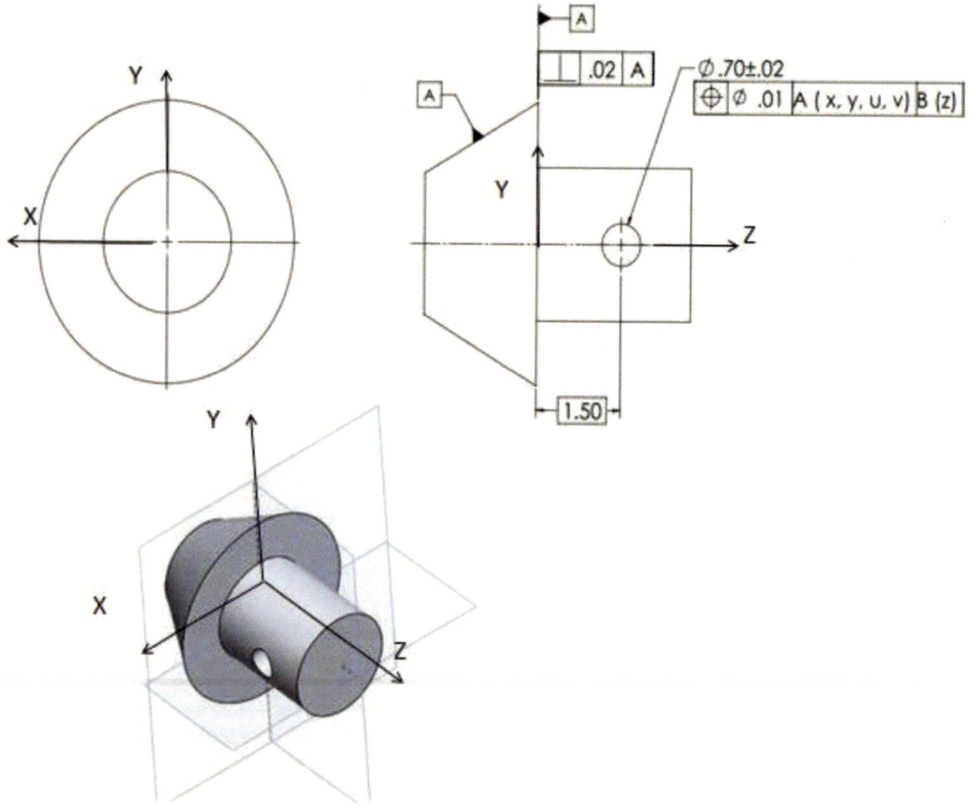

Figure 6.32: Customized Conical Datum Feature

Chapter 7

Tolerance of Location :

7.1 Tolerance of Location – Positional tolerance

Tolerance of location includes positional tolerance, concentricity and symmetry of feature of size. Positional tolerance controls center distance, location and coaxiality of feature of size. It provides a tolerance zone that is located at true position within which axis of produced part must lie.

If it is applied at MMC or LMC condition, it provides a virtual condition boundary that should not be violated. This VC is established at true position.

All the Dimensions for location shall be basic dimension to establish true location. Datum should be established to locate true position except some cases where feature is directly correlated as shown in figure 7.1.

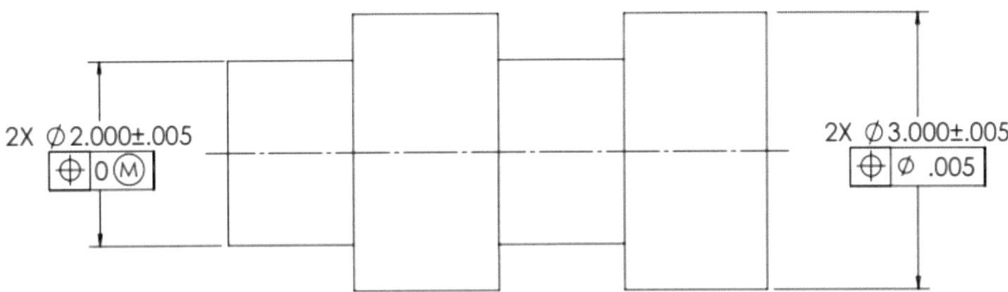

Figure 7.1: Positional tolerance without Datum reference

7.2 Positional Tolerance - RFS

Position tolerance applied regardless of feature size (RFS), when it is required to have axis of unrelated actual mating envelope within designated position tolerance zone established at true position. Axis or center point of feature must be within the zone irrespective of size of feature.

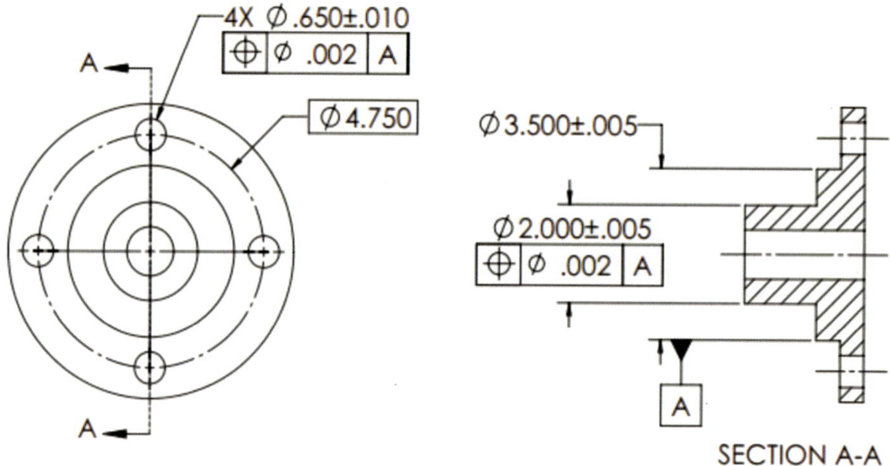

Figure 7.2: RFS

As in figure 7.2 tolerance zone of Ø.002 located at true position with reference to datum A. The feature size ranges from Ø.660 to Ø.640 for pattern hole, and hub diameter Ø2.005 to Ø1.995 should be located within tolerance zone.

Positional tolerance at RFS is more stringent compared if we use MMC or LMC modifier.

7.3 Positional Tolerance – MMC

MMC modifier may be used in feature control frame; it means that if feature is at MMC we get the specified amount of Geometric Tolerance for variation in location. Once it departs from MMC we get the Bonus tolerances.

If feature has extreme form variation (with in feature size limit) or orientation variation of hole, the tolerance in terms of axis may not be exactly equivalent to the tolerance in terms of surface. In those cases surface interpretation takes precedence and Part may accepted even if do not satisfy the axis explanation. (ASME Y14.5 2009 7.3.3.1) see figure 7.3.

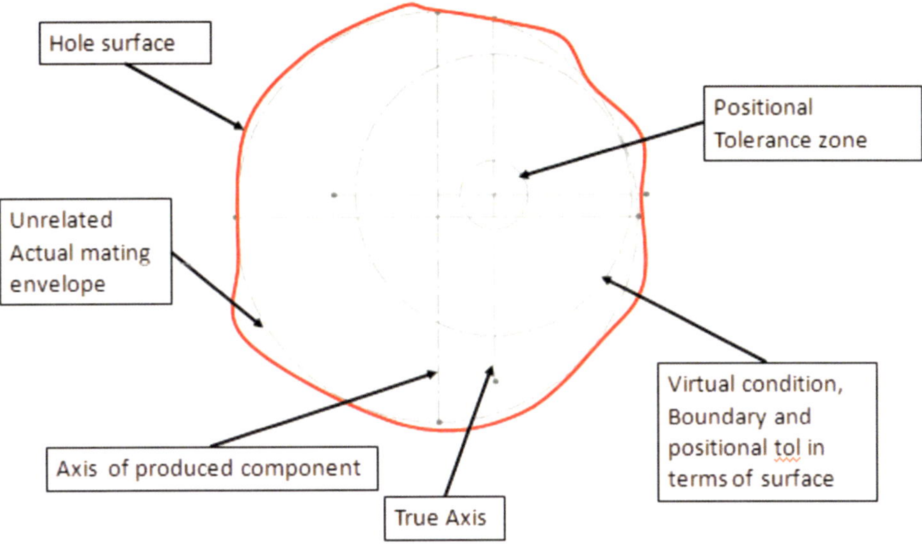

Figure 7.3: Virtual Condition

If produced hole size is within tolerance zone, even if axis of unrelated actual mating envelope is outside of the positional tolerance zone but VC is not violated the part will be accepted.

Surface Interpretation: If component is produced with in size tolerance i.e. unrelated actual mating envelope is with in size tolerance, A boundary (VC) established at true position must not be violated. See figure 7.4.

VC = MMC – Geometric Tolerance (for hole) = .495-.002 = Ø.493 located at true position with basic dimensions .50 and 2.00 must not be violated by actual produced surface otherwise rejected.

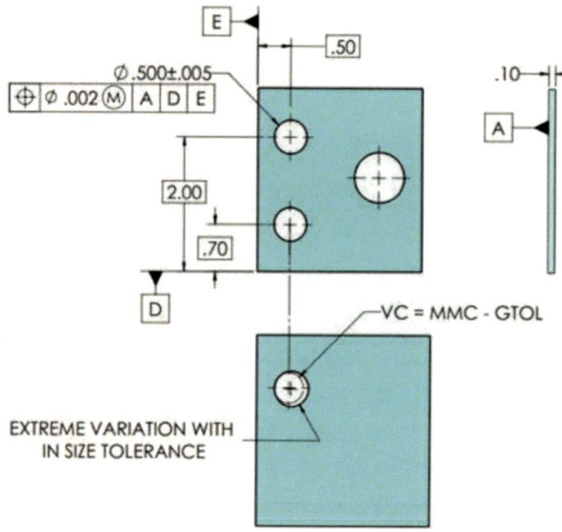

Figure 7.4: Virtual Condition

Now if we think about axis: If the part manufactured at MMC, the features axis must lie within tolerance zone specified located at true position. The orientation of the axis also must fall within the tolerance zone as shown in figure 7.5.

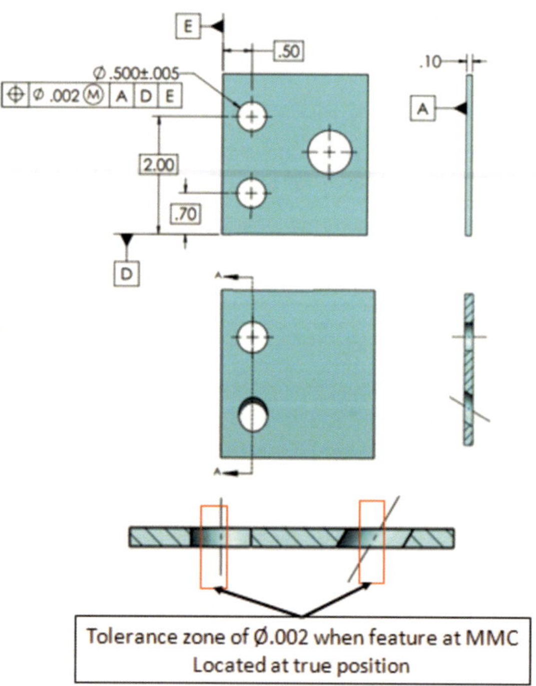

Figure 7.5: At MMC condition

Where actual produced size departs from MMC, we get bonus tolerance. And this Bonus tolerance is equal to the amount of Departure from MMC.

Diameter	Positional Tolerance
LMC = .505	.012
.502	.009
.500	.007
.497	.004
.496	.003
MMC = .495	.002

Table 7.1

As this bonus tolerance does not affect virtual condition, this do not affects functionality or interchangeability of part and it is accepted. This method of tolerance provides additional tolerance to manufacturer without compromising with function.

Floating formula:

If two plates have clearance holes and both to be assembled with Bolt we can calculate tolerance by formula as below:

$T = H - F$

T = Tolerance value

H = Smallest hole size

F = Major diameter of fastener (Largest Diameter)

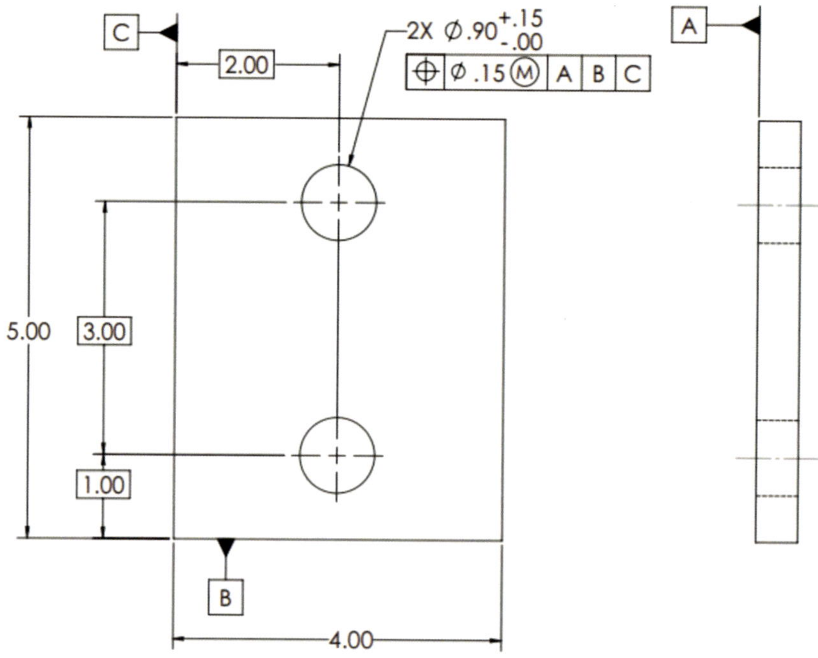

Figure 7.6: At MMC condition

Considering Ø.75 UNC fastener,

$$T = H - F$$
$$T = .90 - .75 = .15$$

This formula do not considers any other variation or error.

Diameter	Positional Tolerance
LMC = 1.05	.30
1	.25
.95	.20
MMC = .90	.15

Table 7.2

The table 7.2 shows the bonus we get with departure of hole from MMC.

Zero at MMC:

When we apply MMC we get additional tolerance as bonus when feature departs from MMC considering it is with in size limit.

However usable part can be get rejected as it is slightly outside the limit of part. Considering example 7.6, Fastener max diameter is Ø.75, suppose manufactured hole is at Ø .85 or even Ø.80 and it is near to the true position it can accommodate part but will be rejected as it fall beyond the size limit.

To counter this, if we selected worst case of fastener diameter as hole diameter and allow no variation at MMC condition that works. This is done by 0 at MMC in feature control frame. This increases size tolerance keeping upper limit of Hole same as Ø1.05 and lower limit to Ø.75.

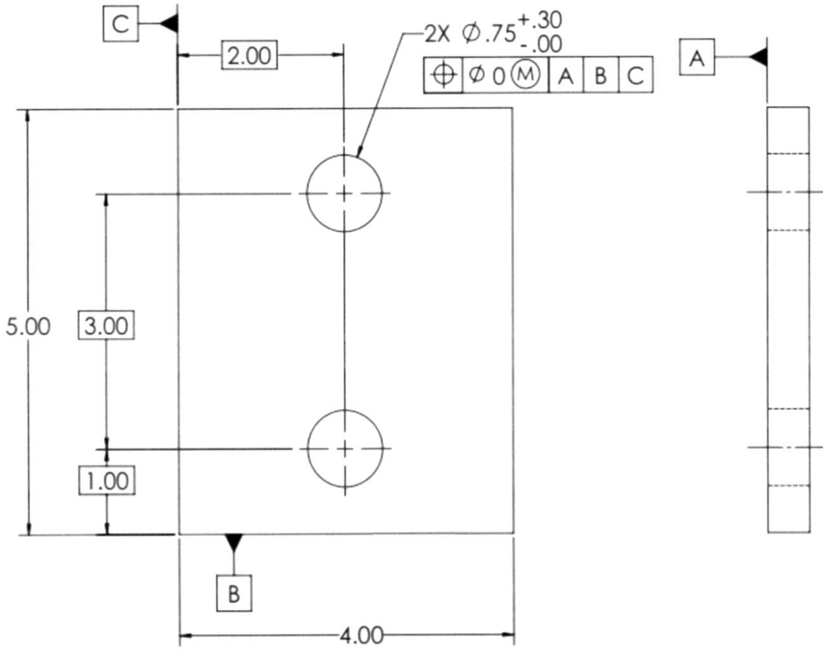

Figure 7.7: Zero at MMC condition

Diameter	Positional Tolerance
LMC = 1.05	.30
1	.25
.95	.20

.90	.15
.8	.05
MMC = .75	0

Table 7.3

If we compare table 7.3 and 7.2 we can understand that the functionality and tolerances remains same but size tolerance got increased, this method communicates design intent perfectly.

7.4 Positional Tolerance – LMC

If we use LMC modifier in feature control frame, LMC condition becomes important to us and the tolerance in feature control frame Applies when feature is manufactured at LMC.

This becomes important when wall thickness or minimum separation distance need to be maintained.

When feature (Unrelated actual mating envelope) departs from LMC we get Bonus tolerance.

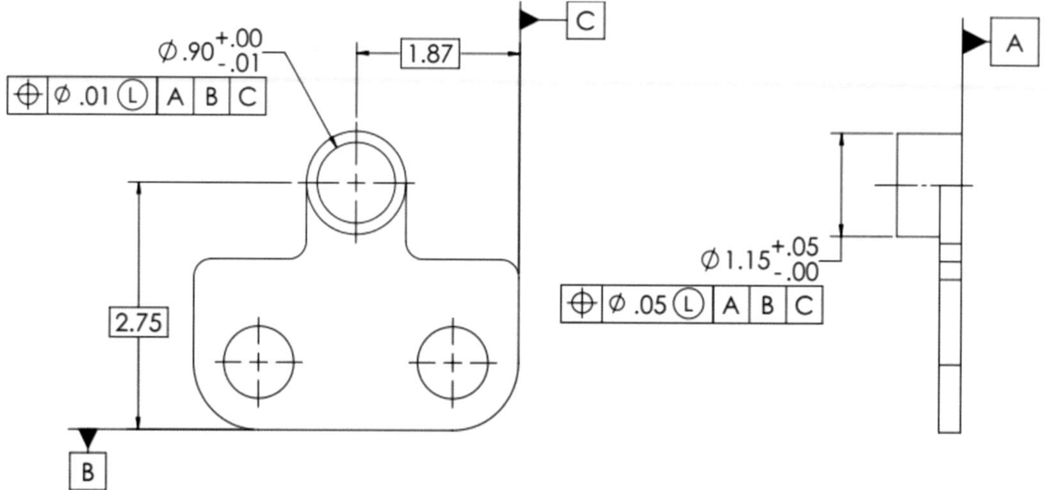

Figure 7.8: LMC condition

To maintain minimum thickness we used LMC as shown. Let see in table 7.4 how it works.

Hole Diameter	Positional Tolerance	Boss Diameter	Positional Tolerance
LMC = .90	.01	LMC = 1.15	.05
.895	.015	1.175	.075
MMC = .890	.02	MMC = 1.20	.10

Table 7.4

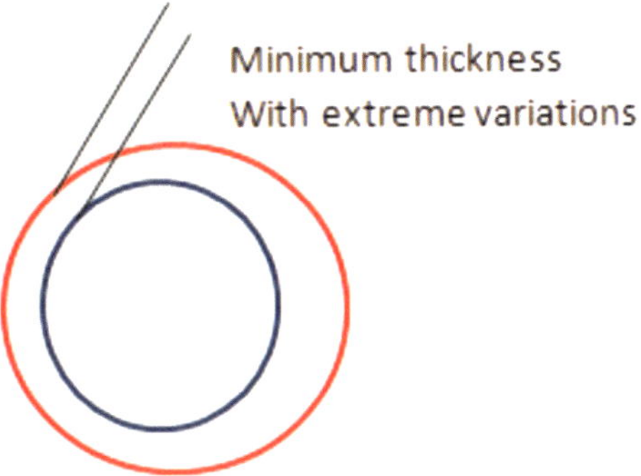

Figure 7.9: Minimum thickness

When Both Features at LMC i.e. Ø .90 and Ø1.15 the tolerance we get is Ø.01 and Ø.05 respectively. So Minimum thickness = ((1.15/2) – (.05)/2) – ((.90/2)+(.01/2) = .55 - .455 = .095 (worst thickness).

When at MMC ((1.20/2)-(.10/2)) - ((.890/2) + (.02)/2) = .55 - .455 = .095

And hence the thickness remains constant with extreme variations. So adding modifier do not changes functionality but eases manufacturing with bonus tolerance.

Zero at LMC:

With LMC application we get bonus tolerances as unrelated actual mating envelope departs from LMC. This is valid when part is with in size limit.

There is a chance of usable component getting rejected. When produced features are at or near true position and slightly away from size tolerance, it may still follow critical thickness requirement but will be rejected.

By using zero at LMC we can make it work as explained in case of Zero at LMC as shown in figure 7.10 and table 7.5

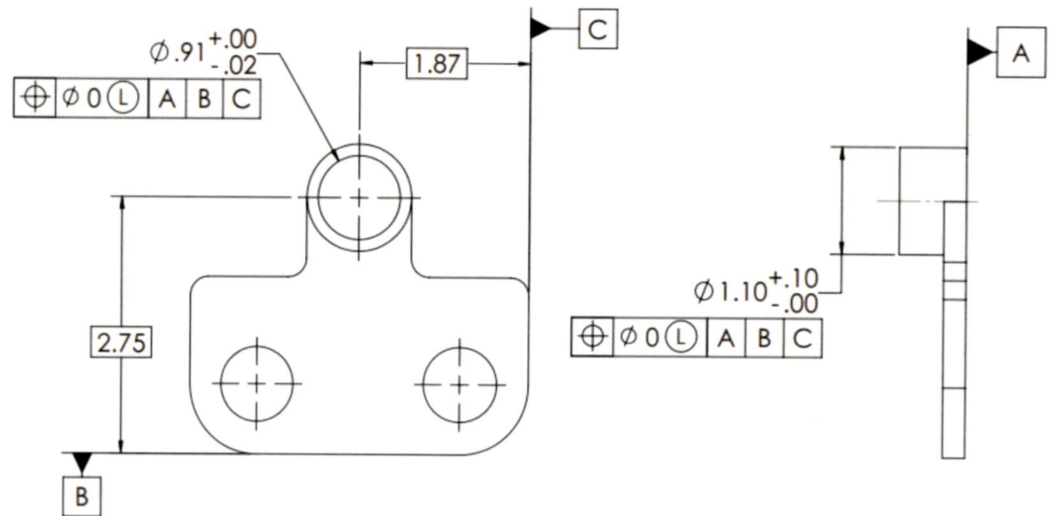

Figure 7.10: 0 at LMC

Hole Diameter	Positional Tolerance	Boss Diameter	Positional Tolerance
LMC = .91	0	LMC = 1.10	0
.905	.005	1.125	.025
.90	.01	LMC = 1.15	.05
.895	.015	1.175	.075
MMC = .890	.02	MMC = 1.20	.10

Table 7.5

If we compare tables 7.5 and 7.4 tolerances are similar and will follow minimum thickness requirement. In this case any location variation we get from amount of departure only.

These terms should not be confused with LMB or MMB. Those applied with datum and provides datum shift. The datum shift should not be considered as a tolerance of location. It is taken care by functional inspection equipments.

7.5 Multiple segment and Composite tolerances:

Feature of size in a pattern may have multiple levels of control. Pattern as a group may have larger tolerance with respect to datum and smaller tolerance within pattern. In those cases we can Composite tolerance or multiple single stage tolerance.

7.6 Composite tolerances:

Composite tolerance symbol has single position tolerance symbol but Feature control frame divided in to two or more segments. Composite tolerance provides location control for feature of size in pattern as a group and it also provide controls of individual feature interrelation within pattern.

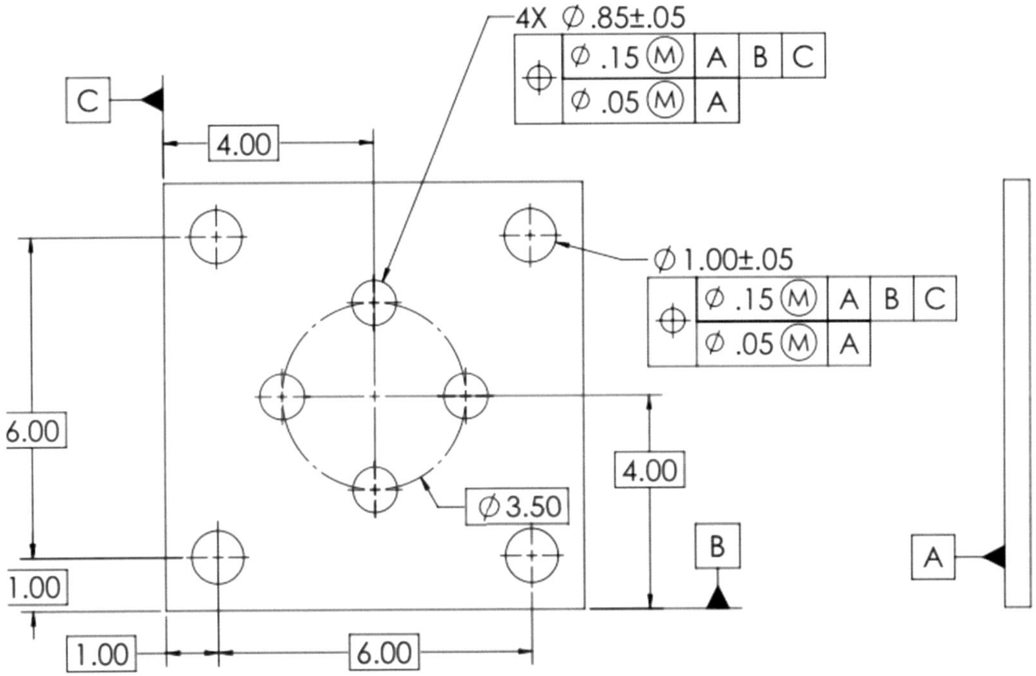

Figure 7.11: Composite Tolerancing

The upper frame in composite tolerance is pattern locating Tolerance zone frame (PLTZF).

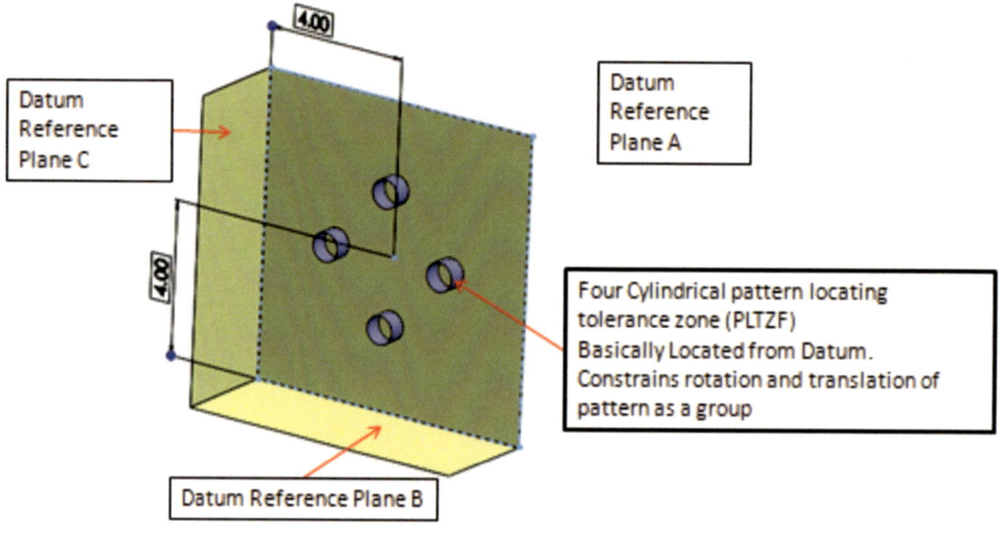

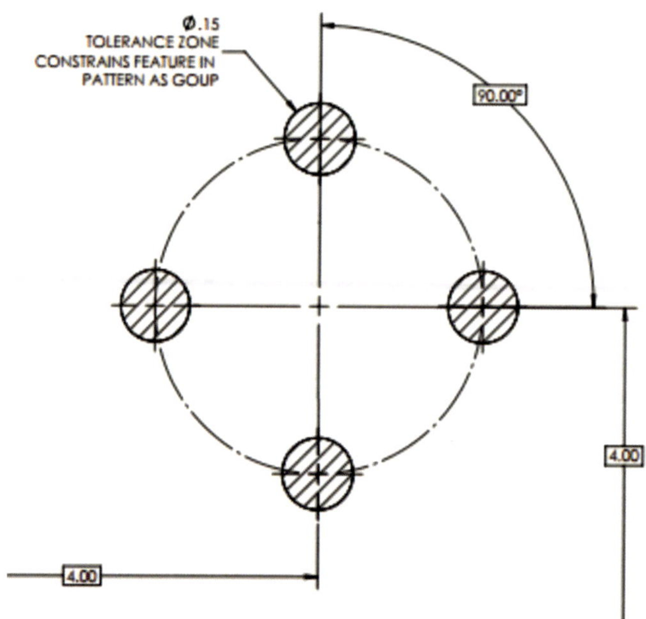

Figure 7.12: Composite Tolerancing - PLTZF

PLTZF tolerance zone is constrained in rotation and well as translation with the datum mentioned in feature control frame.

It specifies larger tolerance for location of pattern as a group. We can say PLTZF is outer boundary established by Geometric Tolerance.

The Lower segment is Feature relating tolerance zone framework FRTZF. This is relatively smaller tolerance zone controls features with in pattern. This cylindrical tolerance zone does not positioned with respect to datum mentioned in Frame; the datum only constrains rotational degree of freedom.

When no datum is mentioned on lower segment, the FRTZF is free to rotate and translate with in boundaries imposed by PLTZF.

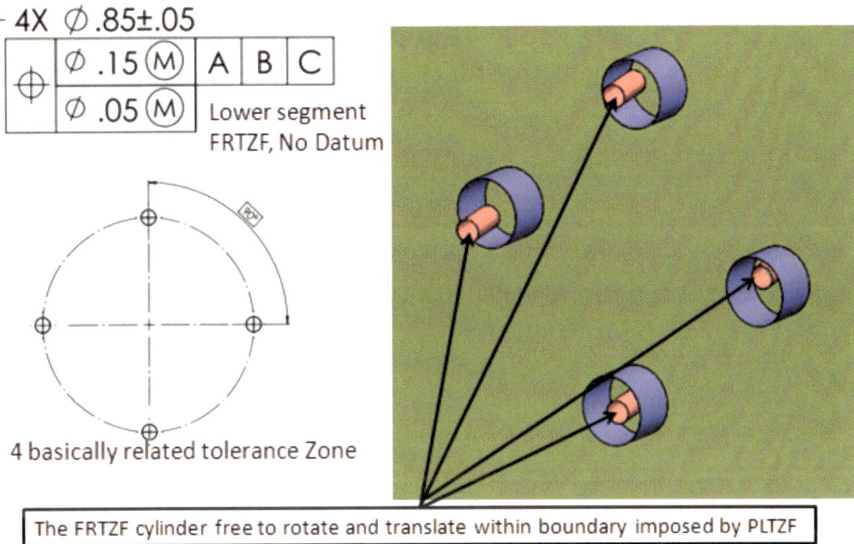

Figure 7.13: Composite Tolerancing – FRTZF without any Datum

If primary datum is repeated in the lower segment, FRTZF and PLTZF both zone oriented with datum A in figure 7.11. These results axes of both cylindrical tolerance zones are parallel to each other. FRTZF can translate and rotate within boundary imposed by PLTZF.

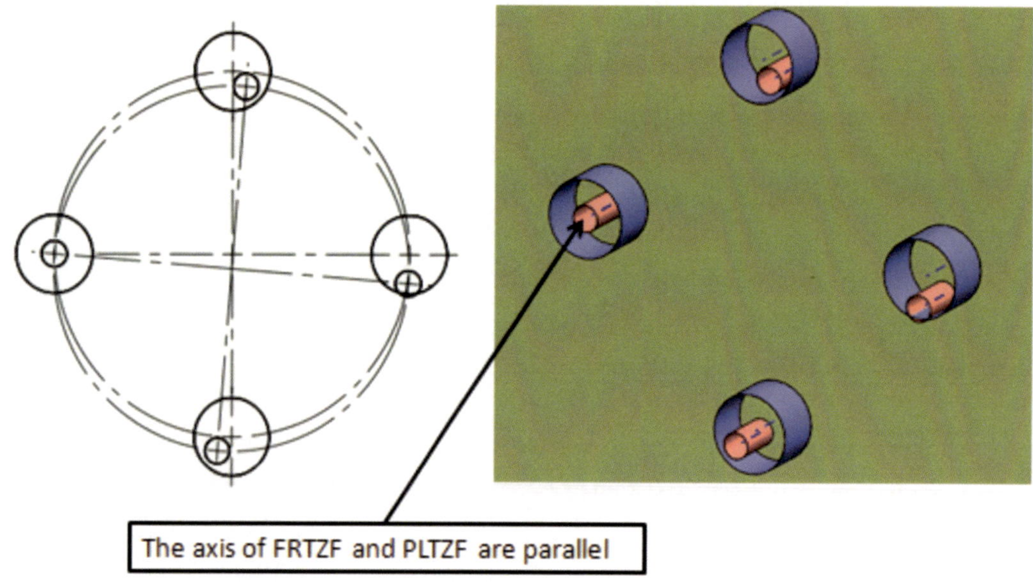

Figure 7.14: Composite Tolerancing – FRTZF with primary datum.

The Tolerance zone FRTZF can go outside of the PLTZF but axis of controlled feature must be within both tolerance zones.

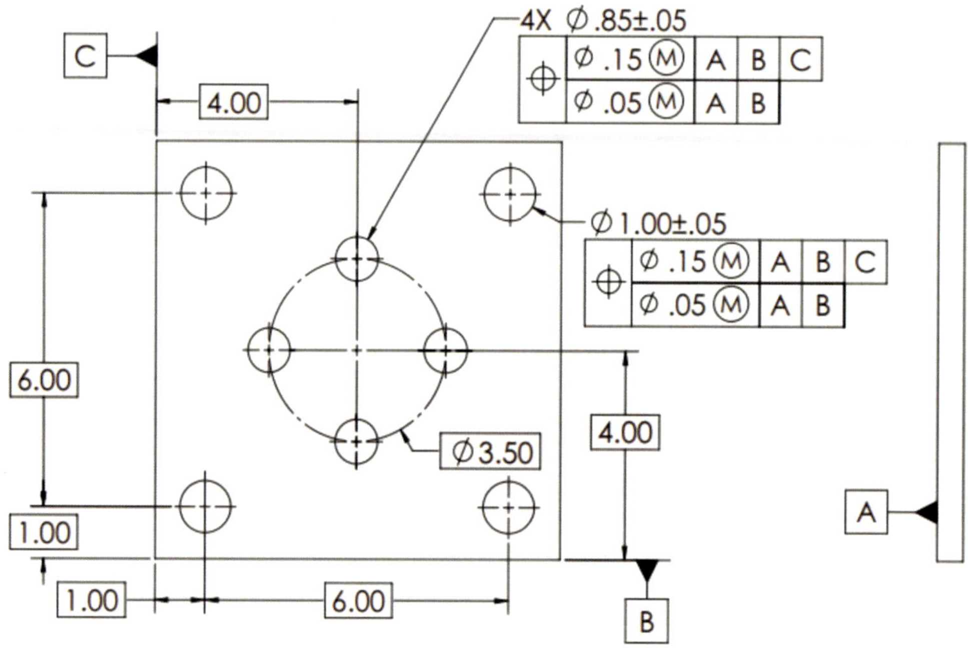

Figure 7.15: Composite Tolerancing – FRTZF with primary datum and secondary datum

If Primary and Secondary datum are repeated in lower segment as shown in figure 7.15, the PLTZF is established as usual located with reference to the datum and constrained in rotation and translation. The four FRTZF cylinder is oriented with datum A and B (rotational constrain) but it can translate within the boundaries established by PLTZF.

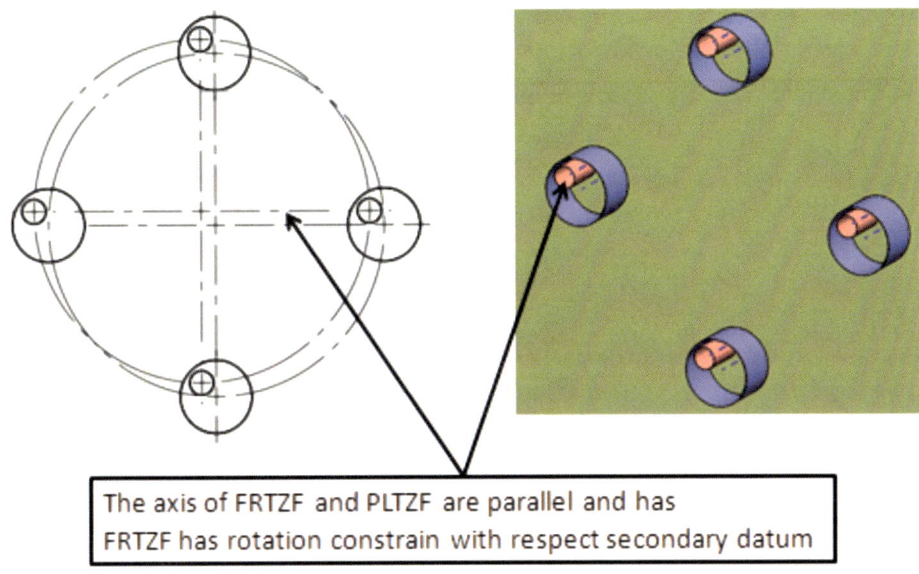

Figure 7.16: Composite Tolerancing – FRTZF with primary datum and secondary datum

Whenever rotational constrain is important we can repeat all datums, Primary, Secondary and tertiary in lower segment.

⊕	⌀ .15 Ⓜ	A	B	C
	⌀ .05 Ⓜ	A	B	C

The FRTZF is perpendicular and parallel to datum A, B and C but it can translate with in PLTZF.

Composite tolerance can be used with radial holes as well. Lengths of these tolerance cylinders are equal to the thickness of the part. There can be multiple lower segments based on requirement.

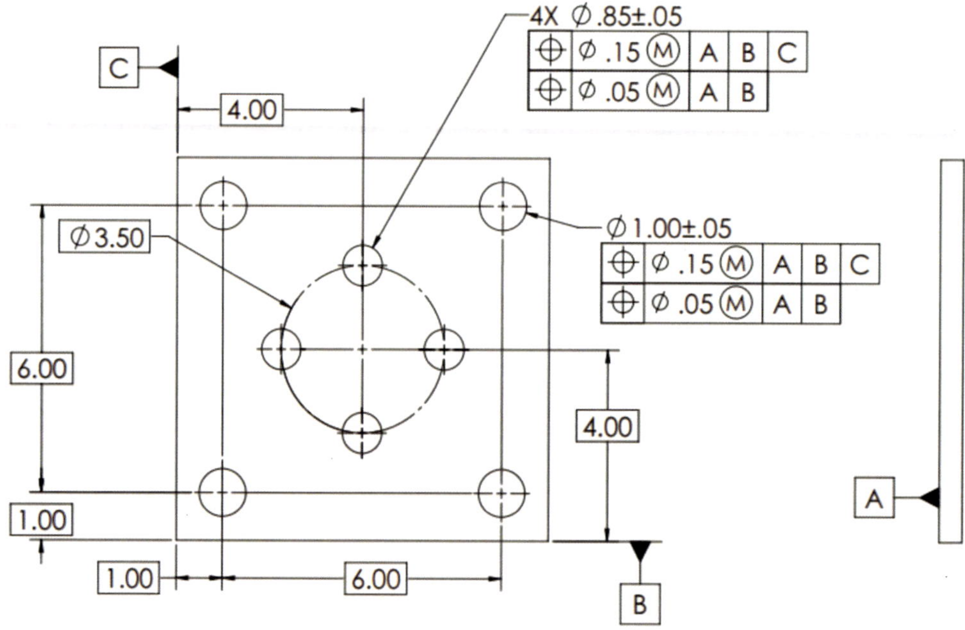

The first segment is PLTZF constrained in rotation and translation w.r.t. Datum A, B and C. The second segment establishes FRTZF and constrained in rotation w.r.t. Datum A and B. The Third segment is also FRTZF without any rotational constrain. Third tolerance cylinder can rotate and translate within the boundaries.

Virtual condition and explanation in terms of surface remains same. Both Virtual conditions should not be violated. And both can be measured separately during inspection.

7.7 Multiple Single Segment Tolerance:

In case of composite tolerance the datum are not used to locate FRTZF and basic dimensions are not used to establish location. When design requirement is such that the basic dimension and location constrain need to be imposed, multiple single segments are used. As shown below:

Figure 7.17: Multiple single segment Tolerance

This do not involves any term such as PLTZF or FRTZF. In this case both requirements should be satisfied separately. The Datums must not be repeated completely in any single segment.

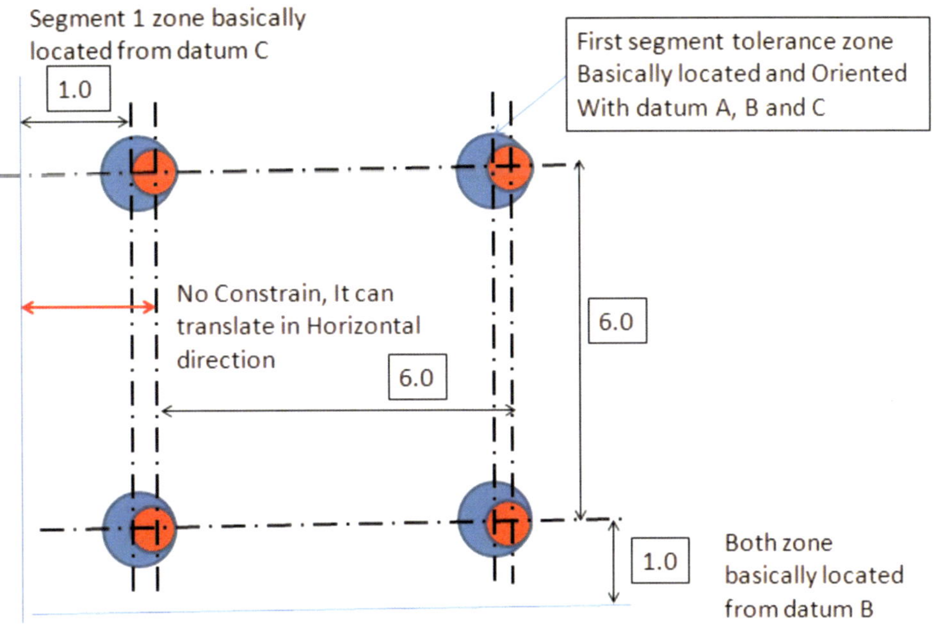

Figure 7.18: Multiple single segment Tolerance

The first segment establishes tolerance zone constrained with reference to datum A, B and C.

Second segment tolerance can move in horizontal direction but within first tolerance zone. If it goes slightly outside that is not used. Axis must be with in both tolerance zones.

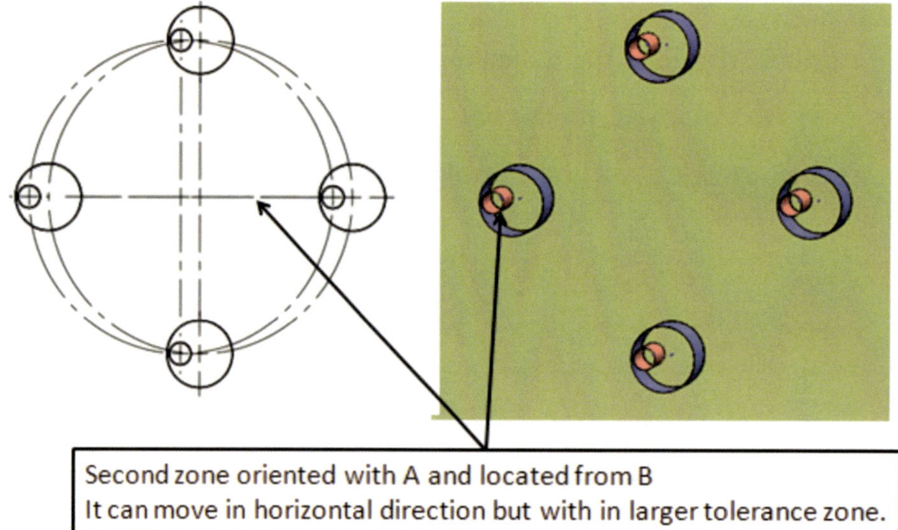

Figure 7.19: Multiple single segment Tolerance

Multiple single segment tolerance zones can be used with cylindrical parts as well. If a datum is an Axis, these results in coaxial tolerance zone segment 1 and segment 2. In that case both zones are coaxial and second tolerance zone can rotate within segment 1 tolerance zone.

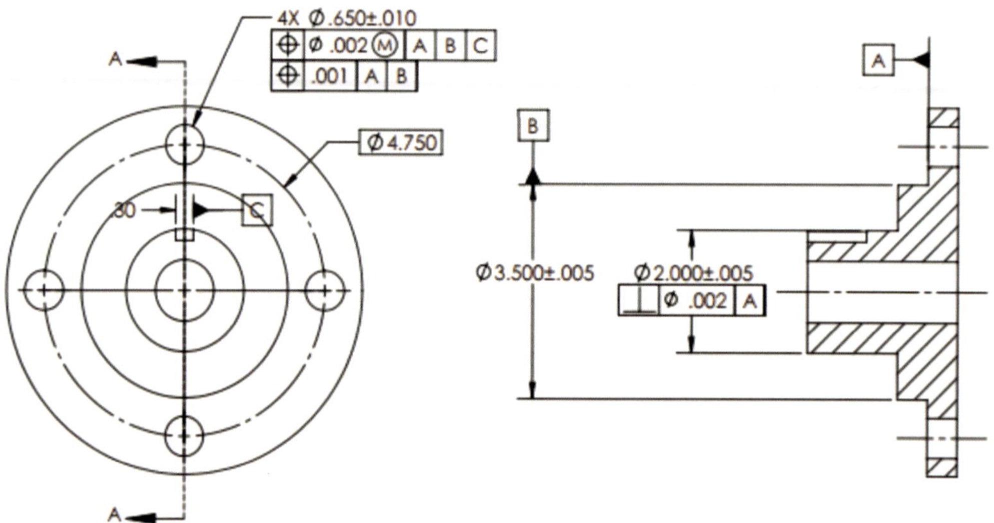

Figure 7.20: Multiple single segment Tolerance

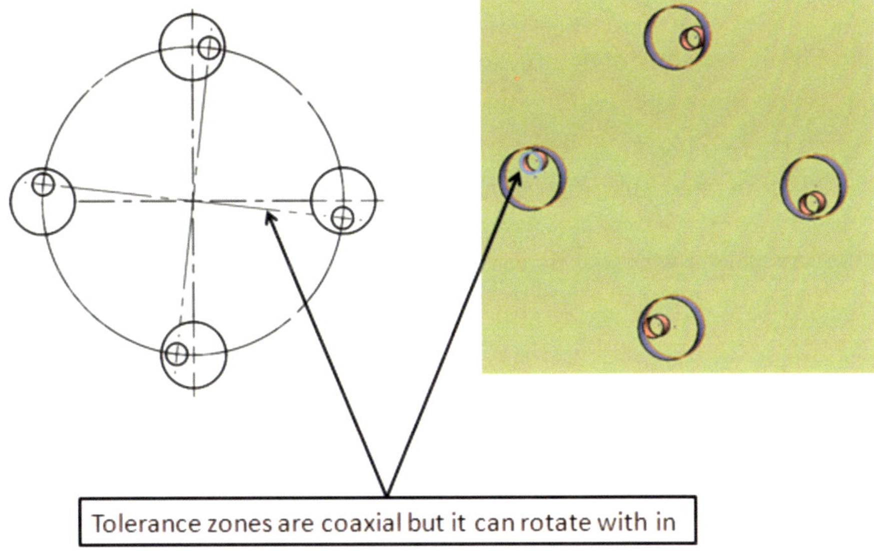

Figure 7.21: Multiple single segment Tolerance

7.8 Application of Projected tolerance zone:

Projected tolerance zone is used when fixed fastener i.e. One plate has clearance hole and other has tapped hole or Dowel pins are in application. In these cases the axis of tapped hole may be within defined tolerance zone but still can interfere with mating part's clearance hole. This is due to fastener projected outside of the tapped hole matters and resulting in interference. This can be avoided by using projected tolerance zones. See Figure 7.22

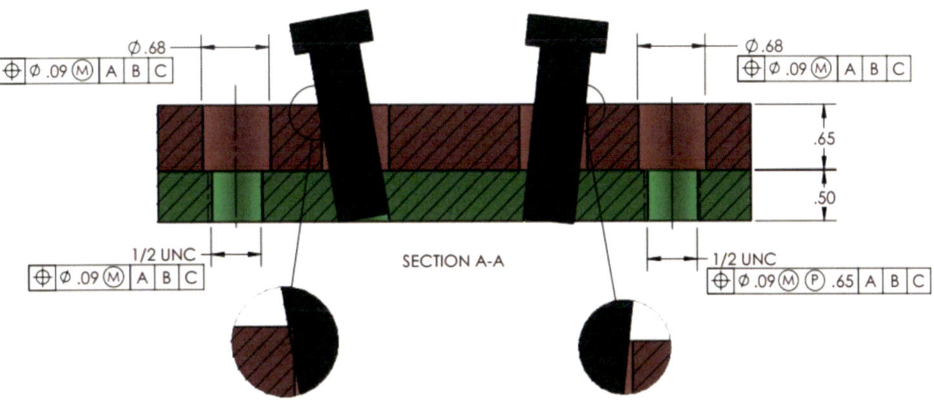

Figure 7.22: Projected Tolerance

Clearance hole of Ø.68 has interference with mating part, whereas use of projected tolerance zone avoids that. In case of floating fasteners i.e. both parts has clearance hole, the error in orientation does not contributes to inclination fastener where as in fixed fastener error causes inclination and interference.

Now by providing projected tolerance, it controls errors of fastener that is outside the tapped hole. Height of projected tolerance zone should be equal to thickness of mating part.

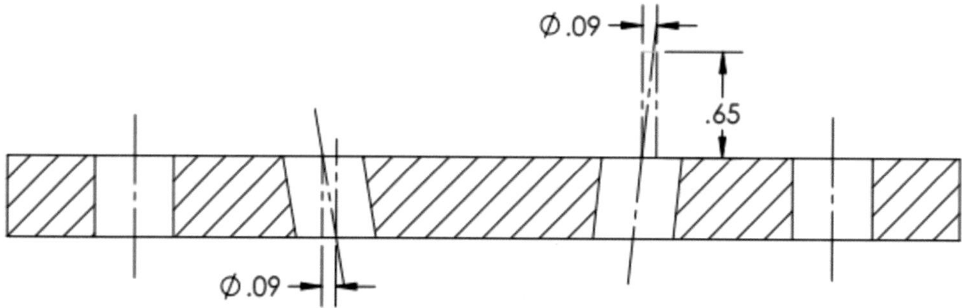

Figure 7.23: Projected Tolerance

Figure 7.23 shows controls and difference between both zones. For projected tolerance we can calculate hole size by following formula:

$$H = D + T1 + T2$$

$$H = \text{Minimum Hole size}$$

$$D = \text{Max Fastener diameter}$$

$$T1 = \text{Positional tolerance on clearance hole.}$$

$$T2 = \text{Positional Tolerance on tapped hole (Projected)}$$

$$H = .50 + .09 + .09$$

$$= .68 \text{ Diameter}$$

If it is fixed fattener and projected tolerance not Applied, to avoid interference Clearance hole diameter should be:

$$H = D + T1 + T2 (1 + 2*P1 / P2)$$

$$H = \text{Minimum Hole size}$$

$$D = \text{Max Fastener diameter}$$

T1 = Positional tolerance on clearance hole.

T2 = Positional Tolerance on tapped hole

P1 = Plate thickness with clearance hole

P2 = Plate thickness with tapped hole.

$$H = .50 + .09 + .09 \, (1 + 2*.65 \, / \, .50)$$
$$= .50 + .09 + .09 * 3.6 = .50 + .09 + .324$$
$$= .914 \text{ diameter to avoid any interference}$$

7.9 Application of Position Tolerance to counter bores, different Positional tolerance at end faces, to non-circular Features:

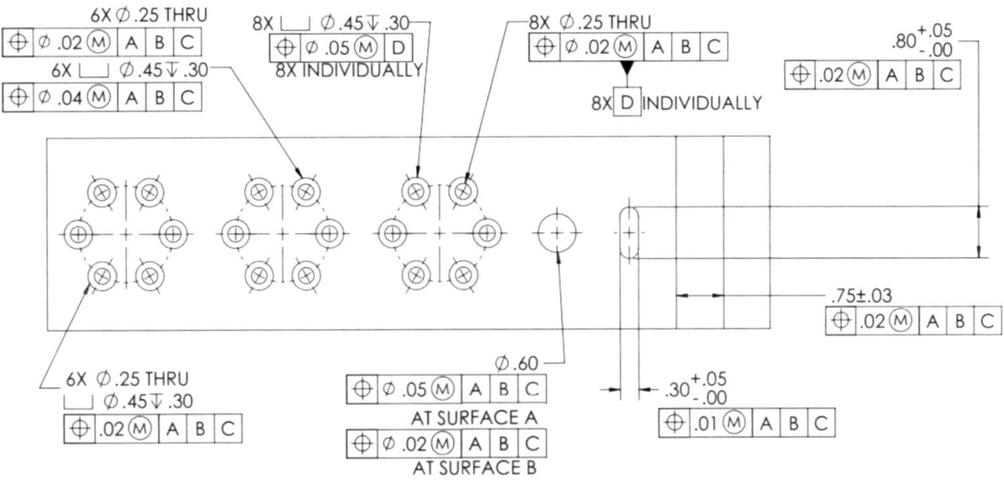

Figure 7.24: Different Applications

A common Geometric Tolerance can be applied to counter bores. In this case single diameter cylinder established coaxially with in which axes can vary from truc position.

If Geometric Tolerance amount is different for through bore and counter diameter, Position tolerance feature frame can be placed separately below both as shown in figure 7.24. Two different cylinder diameters established coaxially and within which unrelated actual mating envelop axis should lie.

Counter bore can be controlled individually w.r.t. datum axis of bore thru, in that case thru bore is made datum and number of feature must be mentioned clearly specifying INDIVIDUAL control.

On two different surfaces as in Figure 7.24 with different tolerance value, Geometric Tolerance can be different this results in conical tolerance zone as shown in figure 7.25.

Position Tolerance can be used on non circular features as slots shown in figure. No diameter symbol is used in this case. It results in a boundary VC that should not violated. And in terms of center plane explanation is similar as cylindrical feature.

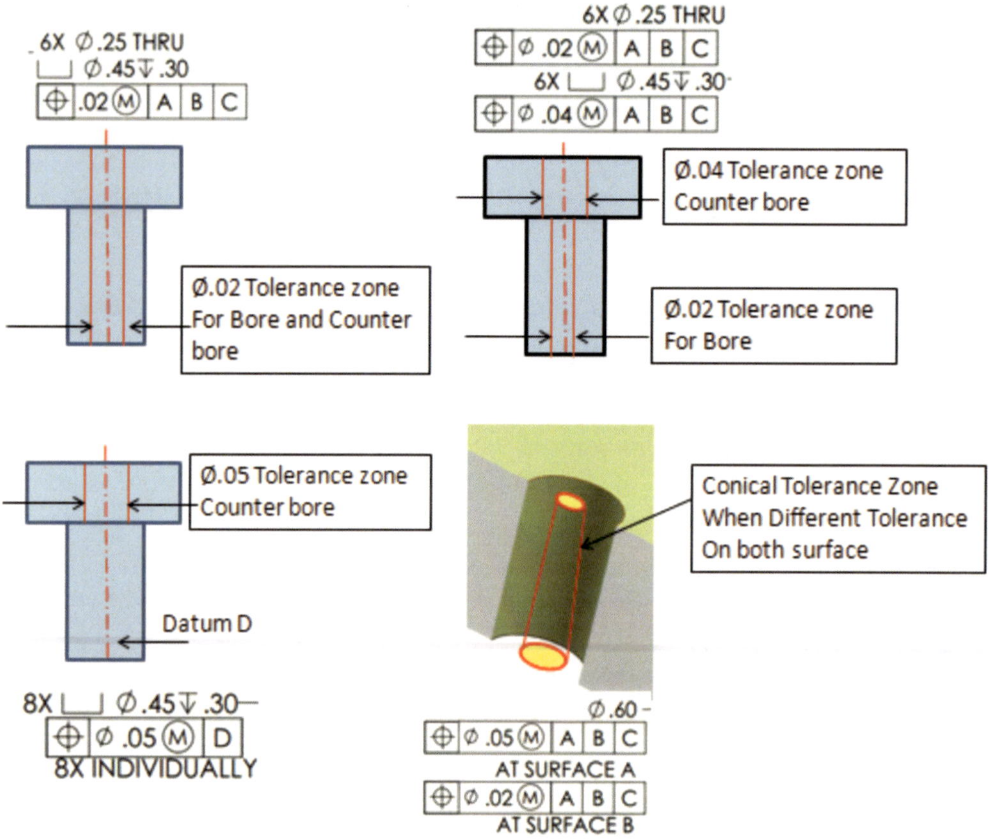

Figure 7.25: Counter bores and Different tolerances at different surfaces

7.10 Coaxial control of pattern feature of size:

Composite Tolerance can be used to control coaxiality between features of size to feature of size and do not restrict excessively for pattern locating tolerance zone.

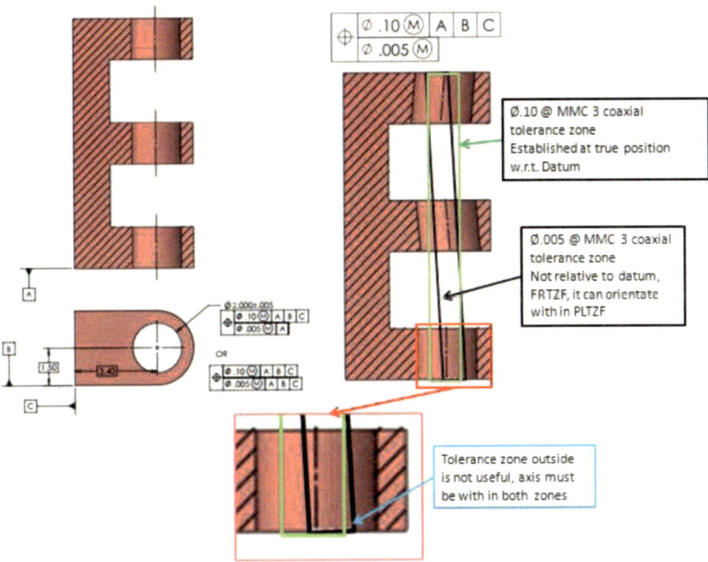

Figure 7.26: Coaxial Control

Same as composite tolerances, first segment establishes PLTZF relative to datums A, B and C. The FRTZF can tilt and rotate and translate but with in boundaries created by PLTZF. PLTZF – three coaxial cylinders located w.r.t. datums. FRTZF – Three coaxial cylinders but not truly located, it can float with in PLTZF.

If we use repeated datums in lower segment the tolerance zone FRTZF parallel and perpendicular to the datums but it can still translate with in PLTZF as shown.

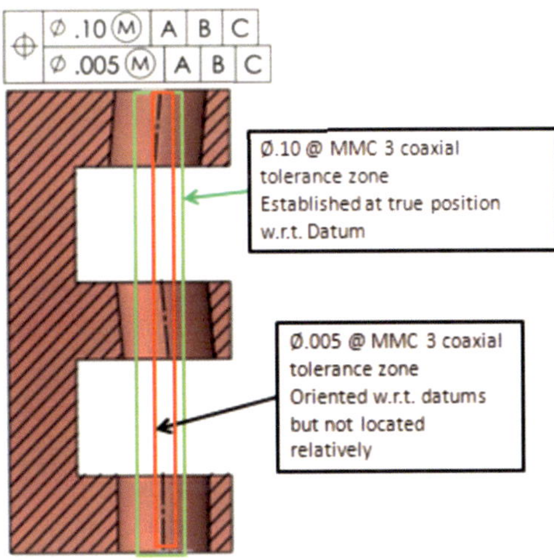

Figure 7.27: Coaxial Control

If sizes of holes depart from MMC each individual feature can get different tolerances depending on amount of departure but the zones will be still coaxial.

7.11 Simultaneous requirement:

When multiple feature sizes are located basically from common datums in same order of precedence at RMB, they are considered as single pattern. There are no relative movements permitted between datum references.

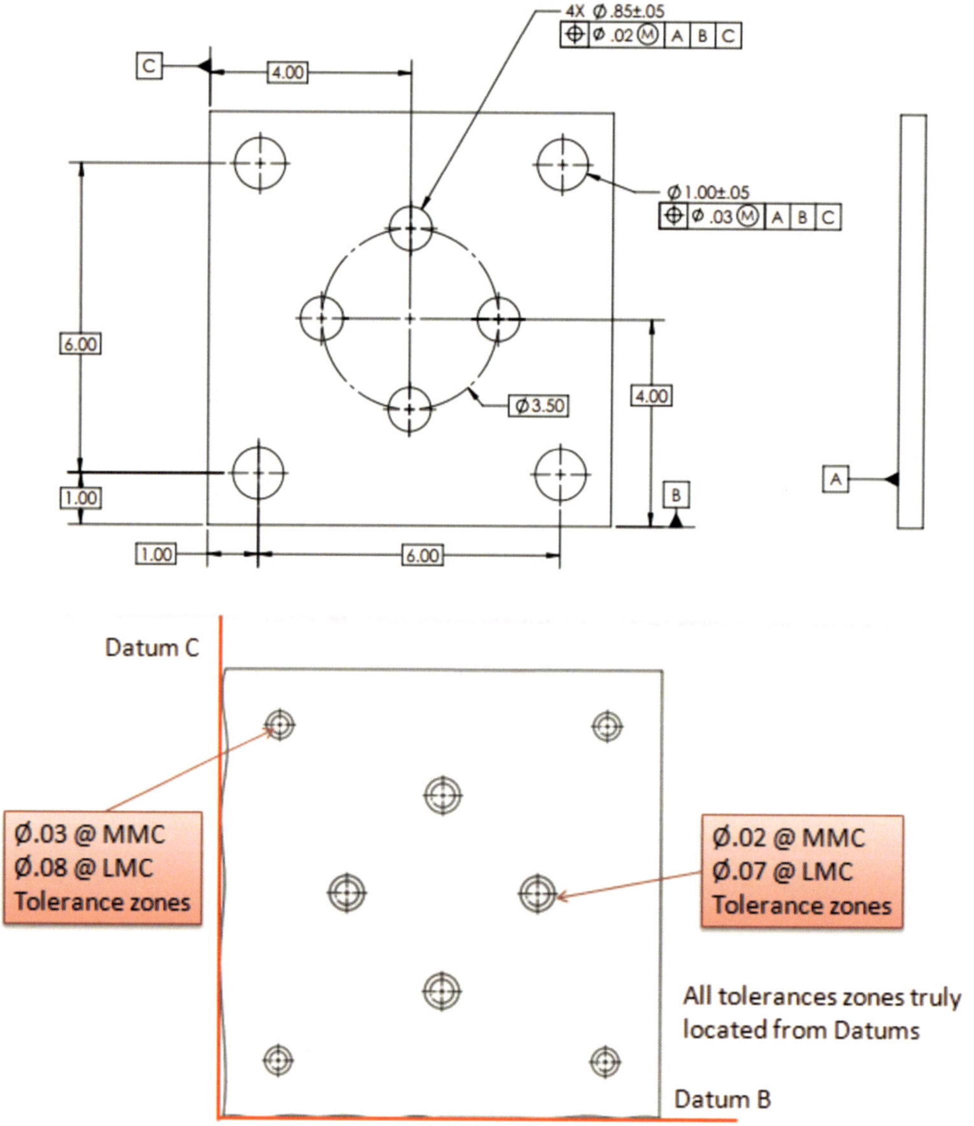

Figure 7.28: Simultaneous Requirement

This is still valid even if measurement done separately due to complex application.

Simultaneous requirement is also possible when datums are mentioned at MMB or LMB. SEP REQ need to be mentioned below feature control frame if the requirement is separate.

7.12 Coaxial Feature of Size:

Coaxiality is axes of actual manufactured feature (Axis of unrelated actual mating envelope or mean points) are coincident with datum Axis. It can be controlled by Position tolerance, profile tolerance, run out and concentricity.

Profile tolerance and Runout discussed in other chapters.

7.13 Positional Co-axial control:

When Axis or surface are controlled and RFS, MMC or LMC and RMB, LMB and MMB conditions are applicable positional tolerance used to control Coaxiality. As shown in figure 7.29.

0 at MMC can be also used, if feature at MMC no coaxiality error permitted and once it departs, we get some tolerance equal to amount of departure.

If Datums at RMB, simulator originates at MMB and moves towards LMB makes max contact and establishes datum axis and boundary is truly located to that and feature is controlled.

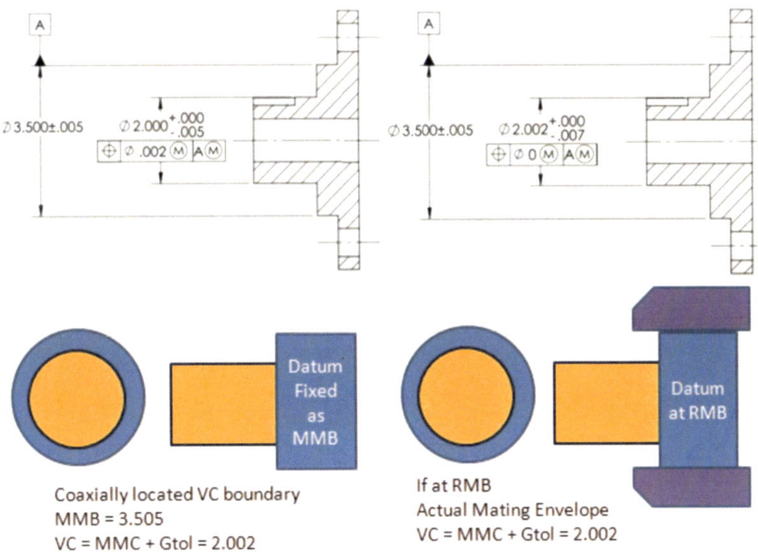

Figure 7.29: Coaxial Control

7.14 Concentricity:

Concentricity is that condition where the median points of all diametrically opposed elements of a surface of revolution (or the median points of correspondingly located elements of two or more radially disposed features) are congruent with a datum axis (or center point). Definition from ASME Y14.5 2009.

The tolerance zone is cylindrical and located coaxially with referenced datum. The derived median points must fall within the tolerance zone. Concentricity application is always at RFS and RMB.

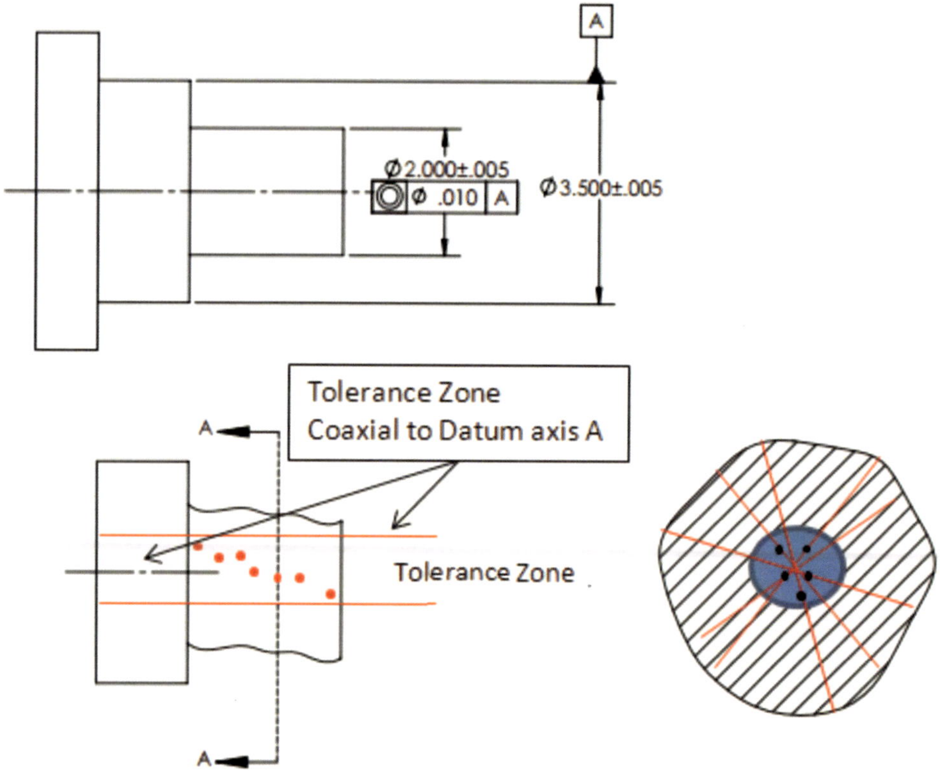

Figure 7.30: Concentricity

7.15 Symmetrical Relationship:

Symmetrical relationship can be controlled by position tolerance, Profile tolerance or Symmetry Tolerance.

7.16 Controlling Symmetrical relationship by Positional Tolerance:

When we control symmetrical relation by positional tolerance, Diameter symbol is not used, The center plane of unrelated actual mating envelope should be congruent with tolerance zone (two parallel planes) established w.r.t. Datum. MMC, LMC or RFS and MMB, LMB or RMB can be used with positional tolerancing. As shown in Figure 7.31.

Datum simulator shows simulation of datum at MMB and Virtual condition boundary truly located at the w.r.t. datum. It is orientated with respect to datum A.

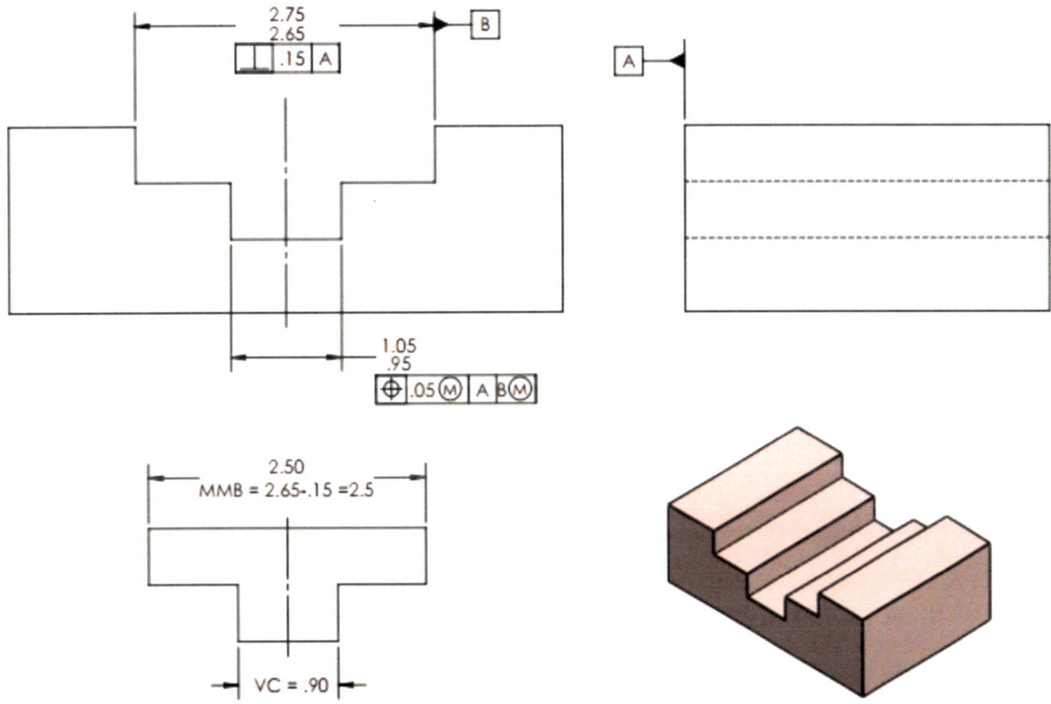

Figure 7.31: Symmetric relationship by Positional tolerance

0 @ MMC positional tolerances can be used in part shown in Figure 7.31. Symmetrical tolerance will depend upon amount of departure from MMC. Figure 7.32 show if it is applied at RFS.

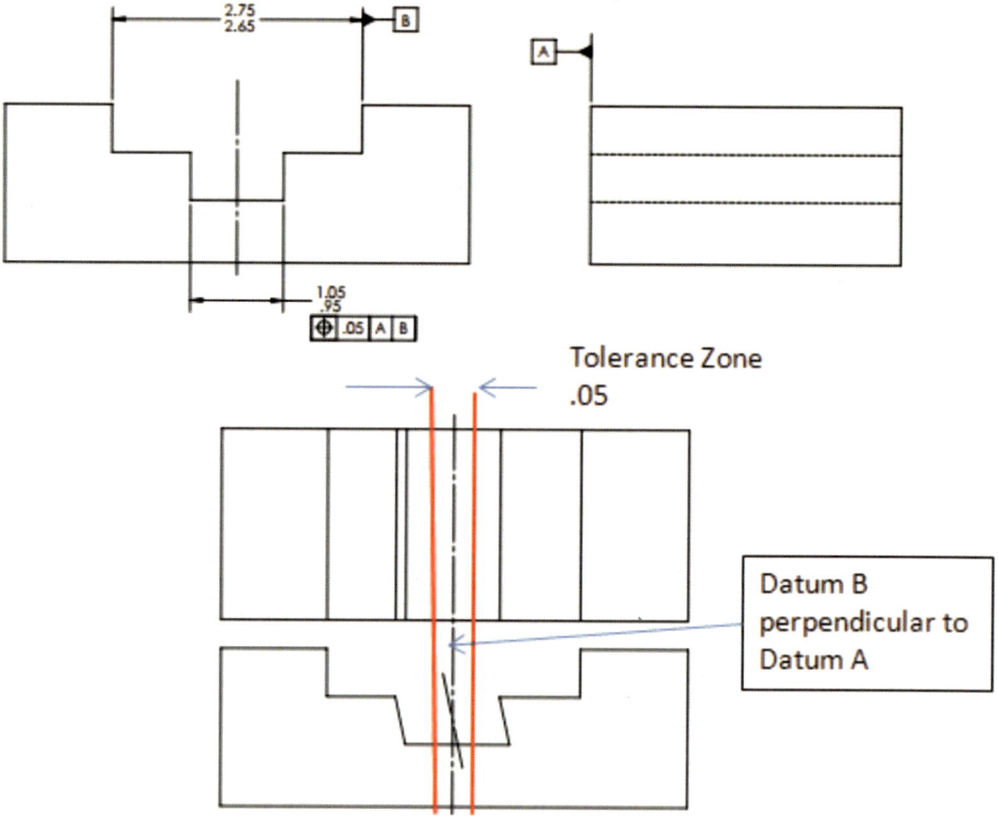

Figure 7.32: Symmetric relationship by Positional tolerance at RFS

7.17 Symmetry Tolerance:

Symmetry is that condition where the median points of all opposed or correspondingly located elements of two or more feature surfaces are congruent with a datum axis or center plane. Definition from ASME Y14.5 2009.

Symmetry is used same as concentricity but for non circular features. It is used on RFS, and RMB basis only.

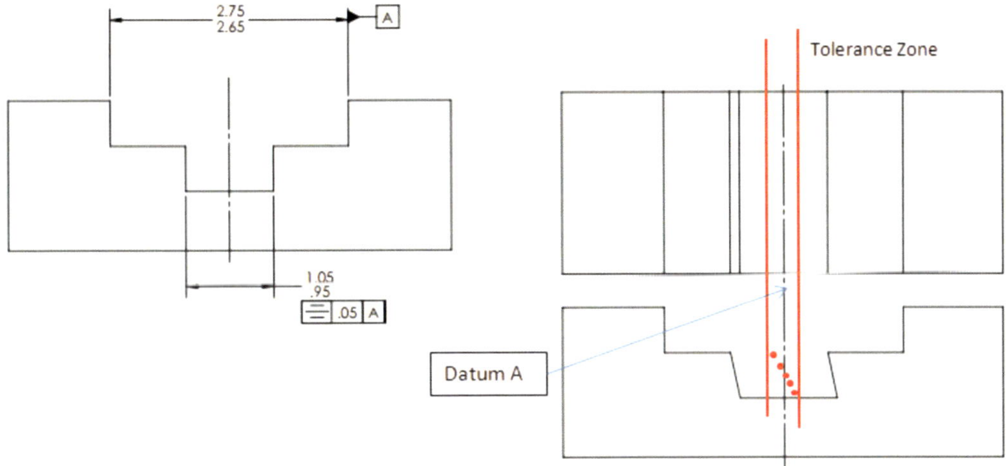

Figure 7.33: Symmetry Control

Median points are derived and must be within tolerance zone located w.r.t. datum reference.

Note: In Latest Revision ASME Y14.5 2018, Concentricity and Symmetry are eliminated due to complexity. But these are still used.

Chapter 8

Tolerance of Profile:

8.1 Profile Tolerance:

Profile tolerance defines tolerance zone that is used to control Form or Size, Form, Orientation and Location depending on requirement. Profile is boundary similar as controlled surface, shape of one or more feature or two dimensional lines.

CAD file is used to define true profile or a drawing view showing all dimensions as basic dimension. When we use profile tolerance with the size tolerance, it should be less then size tolerance for refinement.

	Profile of a Line	RFS	Profile	Individual and Related (With Datum, Without Datum)
	Profile of a Surface	RFS		

Two types of controls are possible with profile tolerance; Profile of a line and Profile of a surface.

Profile of a Surface:

Profile of a Surface tolerance zone is Volume or 3D extends across length and width of feature or features. It can be applied to any shape, constant cross section part or surface of revolution.

Profile of a Line:

It is kind of 2D control, tolerance zone provides area disposed about true profile within which each element of surface must lie. The zone is normal to the profile. It can be applied to any varying cross section or constant cross section.

Profile tolerance can be applied a general note on drawing. Profile tolerance can be related to datums as well depending upon requirement.

8.2 Uniform Profile tolerance zone:

In this case the tolerance zone is uniformly distributed along true profile. The can be equally or unequally or unilaterally disposed to true profile. Profile tolerance is always normal to all points of true profile. Boundaries follow the true profile shape all across the surface.

The surface lies within the defined tolerance zone. Some time on corners, it can produce abrupt surfaces so proper fillet or radius defined in truc profilc.

8.3 Bilateral Tolerance:

In case of equal bilateral tolerance, the zone is equally divided both side of true profile. For this the leader should be directed to the surface or its extension not to the basic dimension.

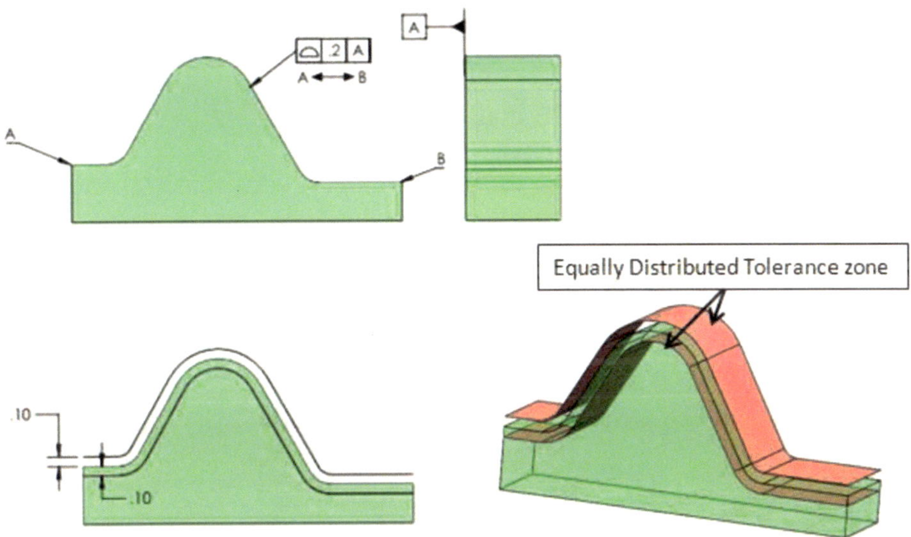

Figure 8.1: Equal Bilateral

8.4 Unilateral and Unequal Bilateral:

To show tolerance is unequally disposed or unilaterally at one side only unequal symbol is used as shown in figures.

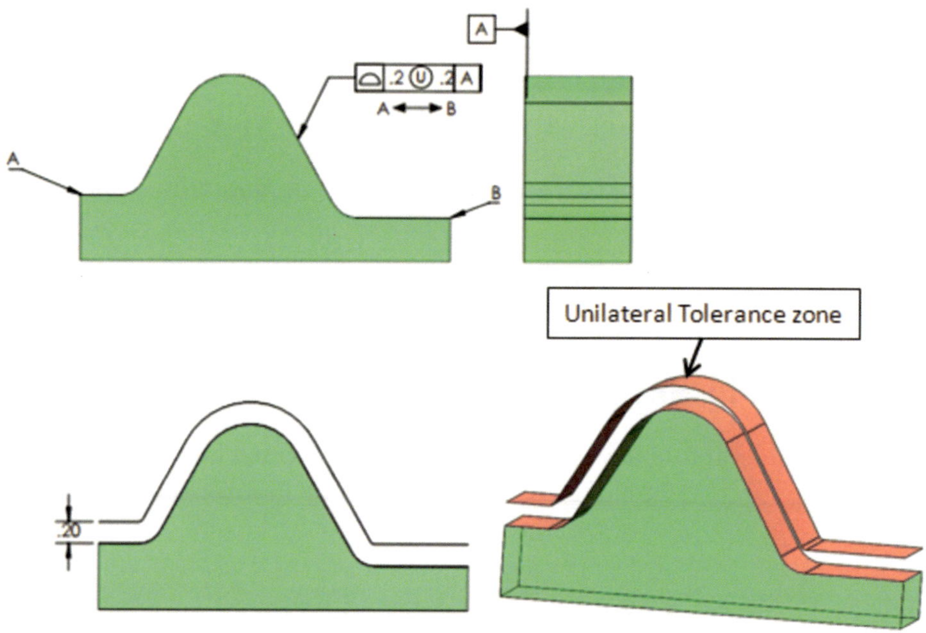

Figure 8.2: Unilateral adding material

If inequality symbol repeats complete tolerance value, it adds material and tolerance is one side only as shown in figure 8.2.

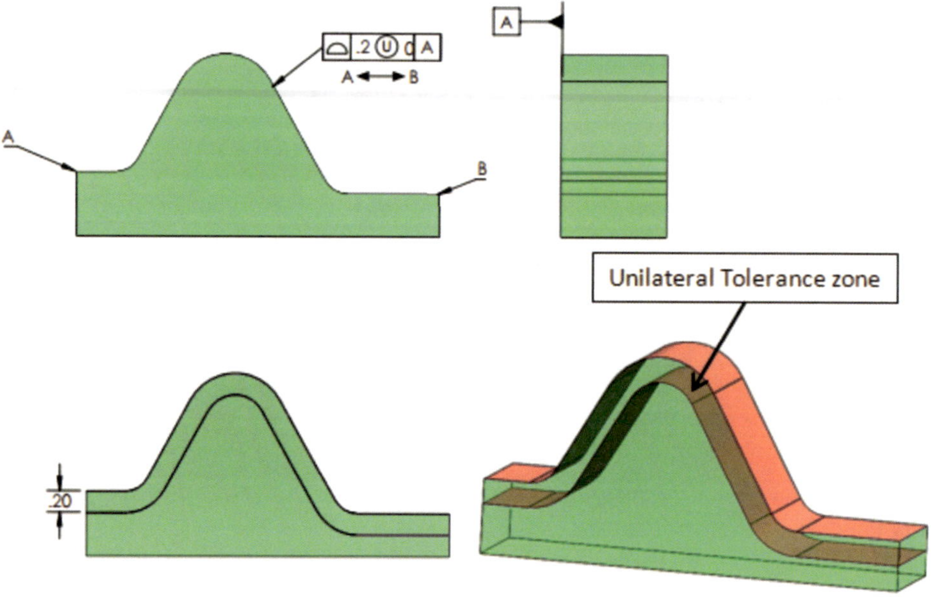

Figure 8.3: Unilateral removing material

If inequality symbol repeats 0 after it, it removes material and tolerance is one side only as shown in figure 8.3.

If requirement is unequally disposed tolerance, it is as shown in figure.

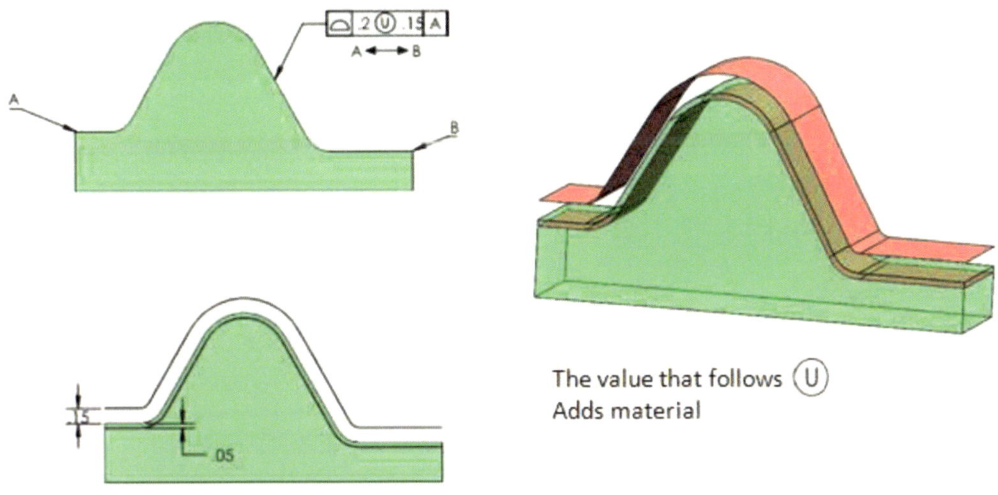

Figure 8.4: Unequal Bilateral adding more material

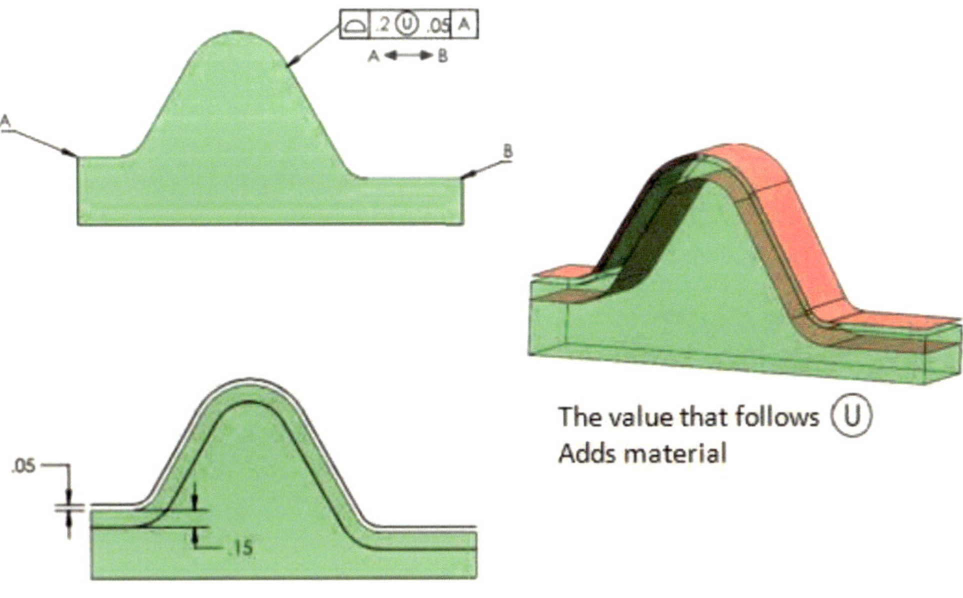

Figure 8.5: Unequal Bilateral removing more material

8.5 All around and All Over symbols:

Where Profile tolerance applies all over the true profile in a drawing view where it is specified, all around symbol is used. It should not be applied in isometric views.

When Profile tolerance is applicable to the all over the surfaces of part All over symbol is used or ALL OVER note is mentioned.

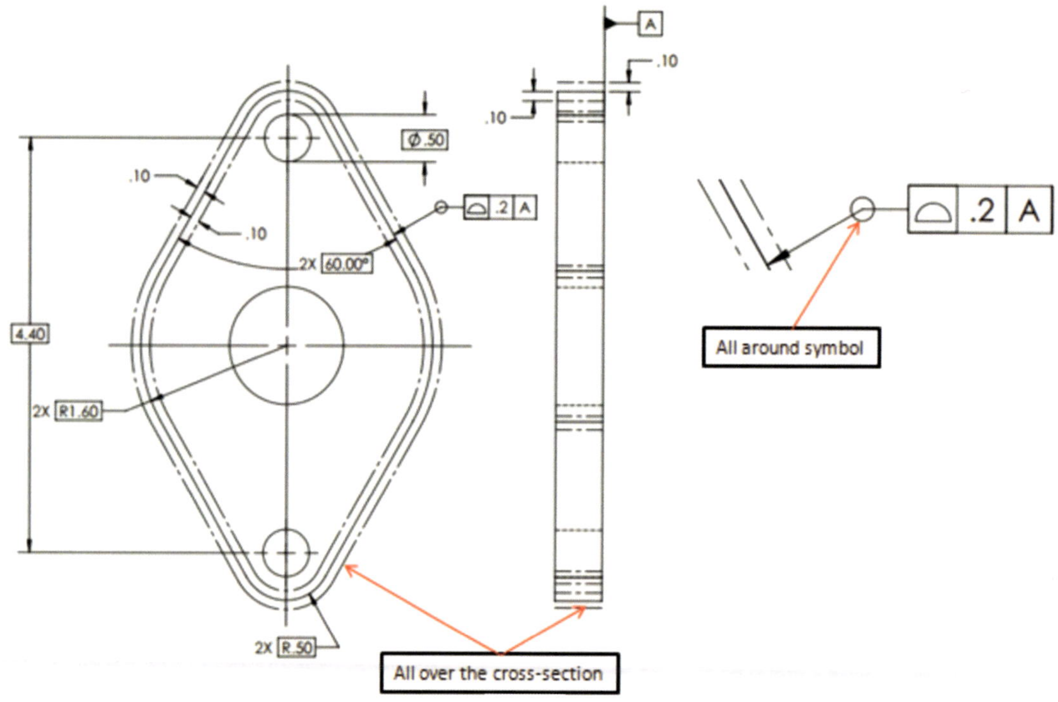

Figure 8.6: All around Application

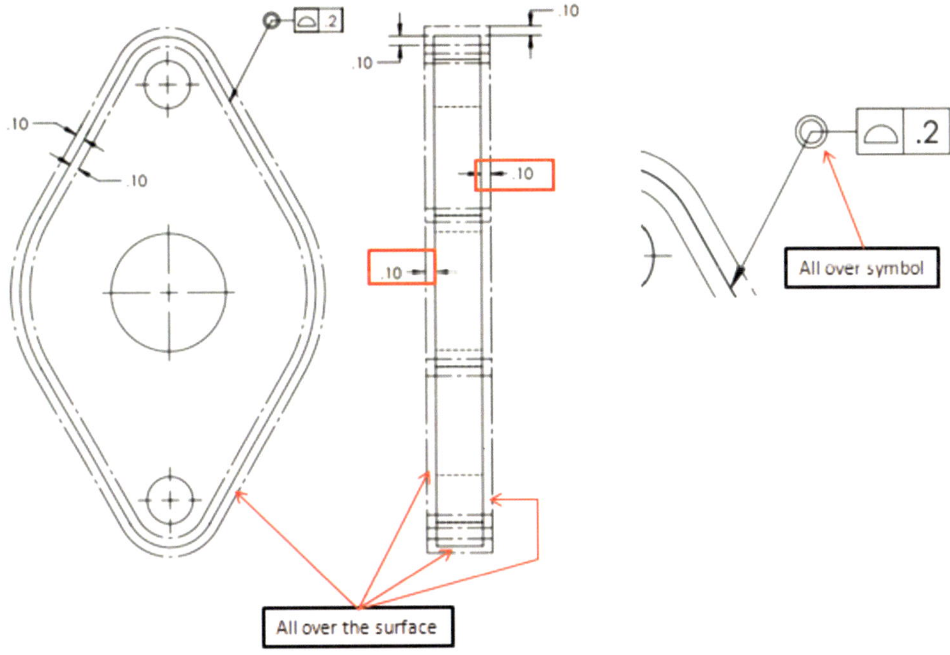

Figure 8.7: All over Application

8.6 Different Tolerance for different segment of cross section:

When different portion needs different profile tolerance zone between symbols can be used and letters can be used to denote boundaries.

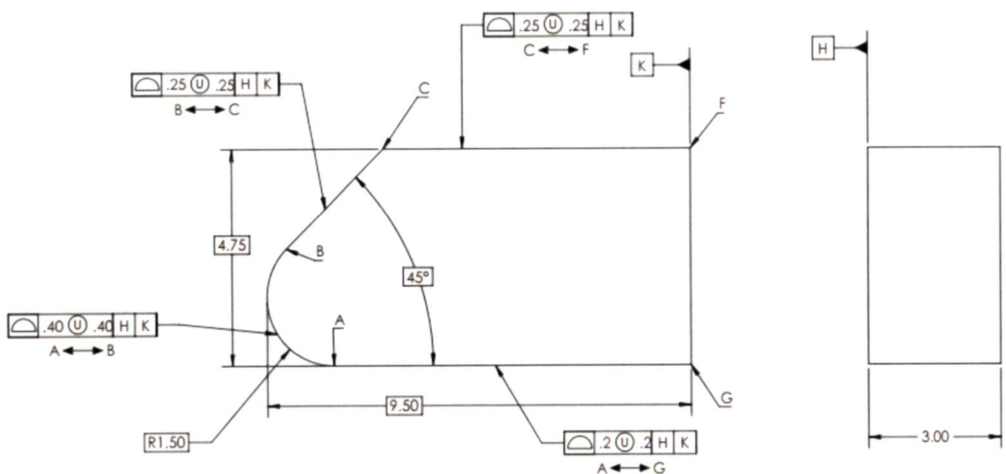

Figure 8.8: Between Symbol and Segmented tolerance

The tolerance zones are different of different segment AB, BC, CF and AG as shown in Figure 8.9.

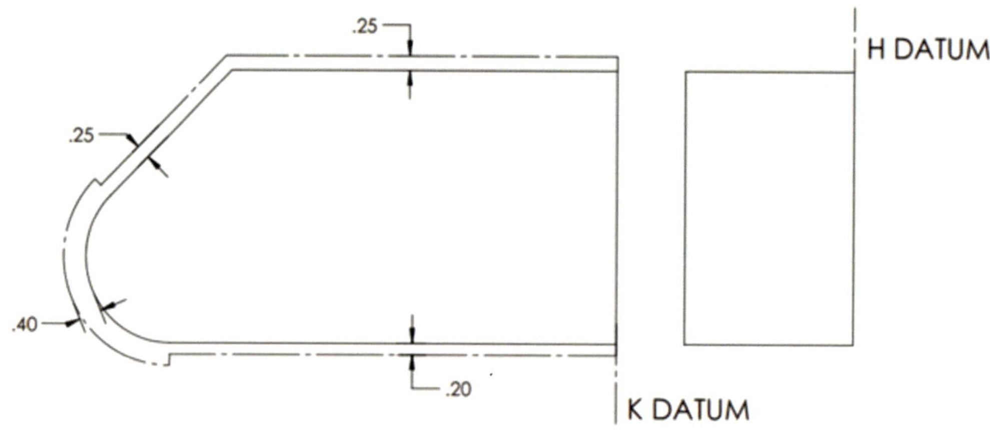

Figure 8.9: Tolerance zones for part in 8.8

8.7 Non Uniform Tolerance Zones:

The boundaries are defined on drawing itself and tolerance value is replaced by NON UNIFORM note. Basic dimension used to define these boundaries.

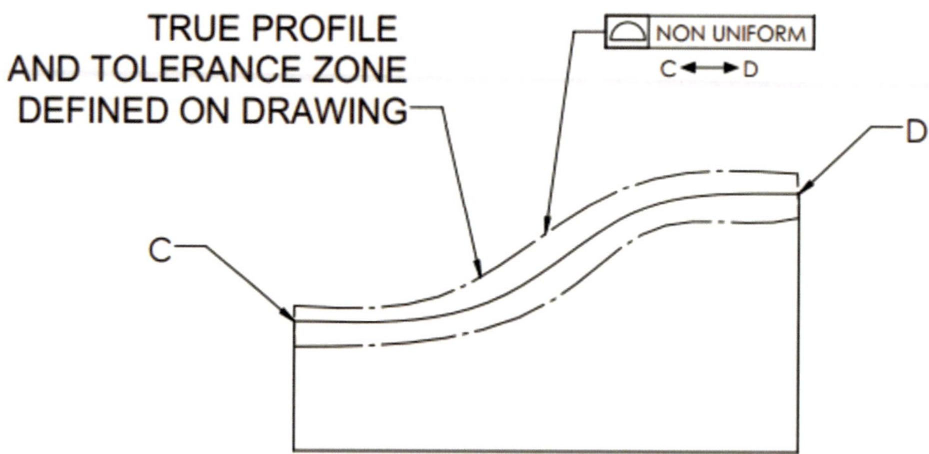

Figure 8.10: Non Uniform Tolerance

This can be used to smoothen out the tolerances. Figure 8.9 shows part has abrupt change in tolerance that can be made smooth by using non uniform tolerance zone as shown.

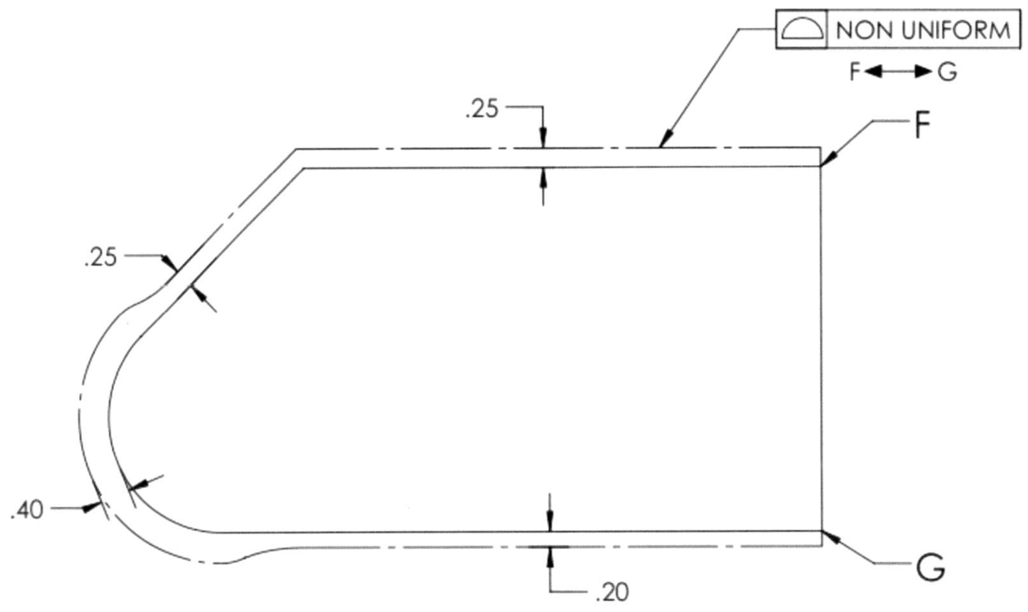

Figure 8.11: Non Uniform Tolerance

8.8 Composite Tolerance Application:

When it is required to control form, size and orientation with fine tolerance and Location of profiled feature or Pattern of feature can have loose tolerance, we find application of composite tolerancing with Profile tolerance. Analogy is similar to the Composite Positional Tolerance.

8.9 Composite Tolerance Application to single feature:

Figure 8.12 shows application of composite profile tolerance. The feature control frame upper segment is feature locating tolerance zone. The Datums in first segment locates profile locating tolerance zone.

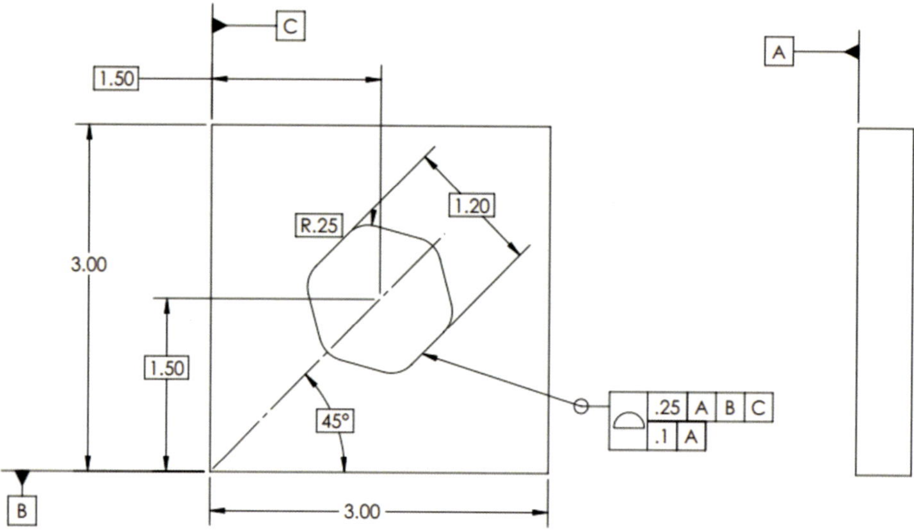

Figure 8.12: Composite Applications

Lower segment controls size, form and orientation of profiled feature. Datum referenced in lower segment establishes limit of size form and orientation relatively to referenced datums. Tolerance value is disposed about true profile as we earlier discussed for unilateral or bilateral application. Actual Surface must lie within both tolerance zones.

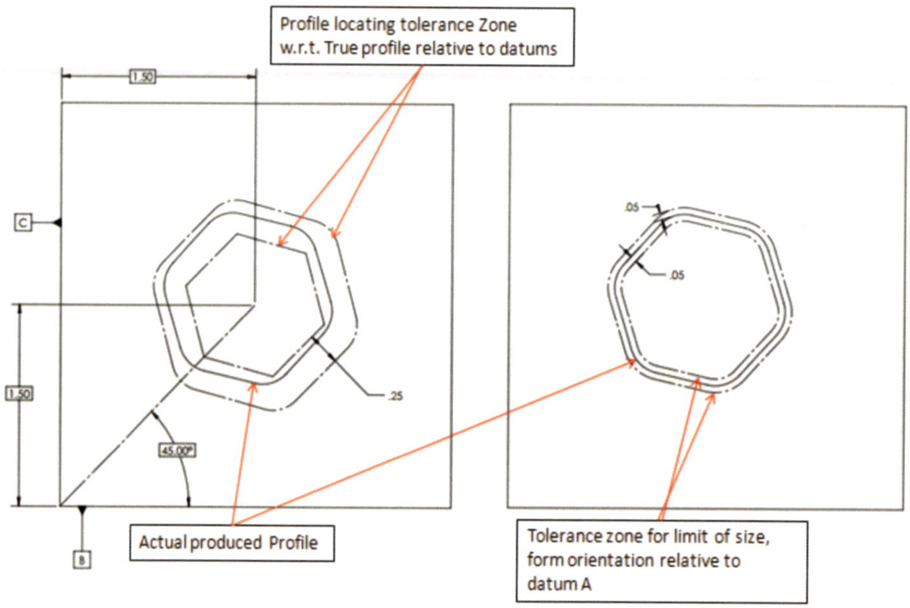

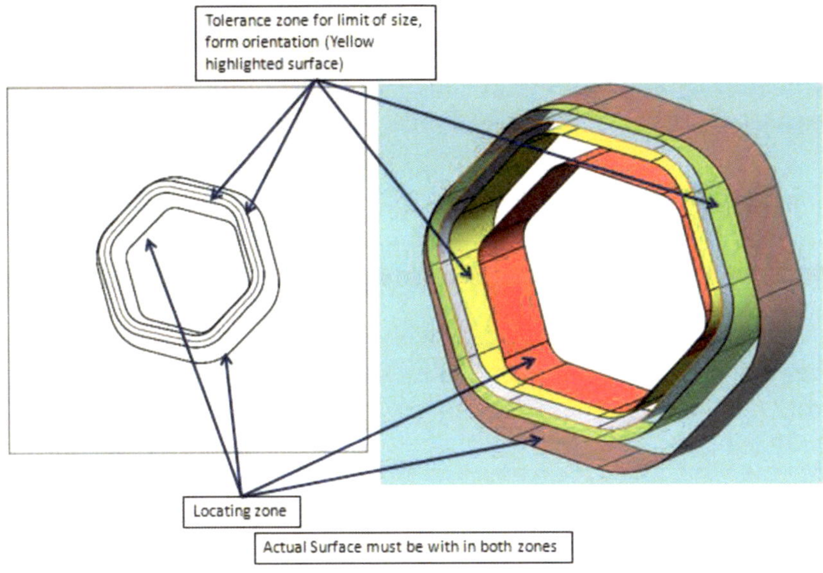

Figure 8.13: Composite Applications

8.10 Composite Tolerance Application for feature in pattern:

Depending upon requirements we can use composite profile tolerance to Pattern of feature of sizes for profiled feature. It states larger tolerance for PLTZF and tighter tolerance for FRTZF. FRTZF can rotate and translate with in boundaries by PLTZF.

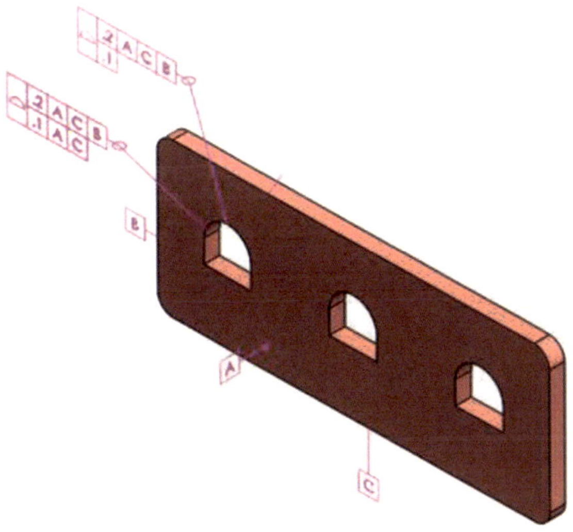

Figure 8.14: Composite Applications

PLTZF: The upper segment is pattern locating tolerance zone. Pattern location tolerance zone is established relative to datum in order of precedence. It is constrained in rotation and translation. It provides outer boundary and has larger tolerance and controls pattern as a group see figure 8.15.

FRTZF: The lower segment is Feature relating control. It is smaller tolerance controls size form and orientation and location for features with in pattern. If any datum in feature control frame, they constrain rotation degree of freedom of FRTZF w.r.t datums.

The actual feature produced must lie with in both tolerance zones. FRTZF is established as shown in figure 8.16. It does not use dimensions 4.00 and 3.00 from Datum C and B. The interrelation basic dimension 5.00 is used. So all three boundaries are basically related to each other.

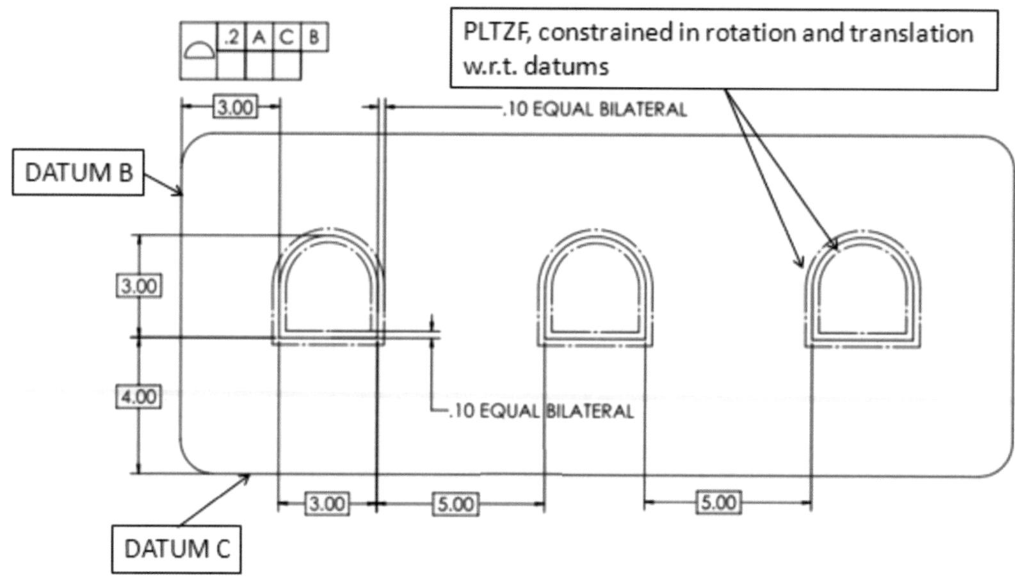

Figure 8.15: Pattern Locating Tolerance Zone

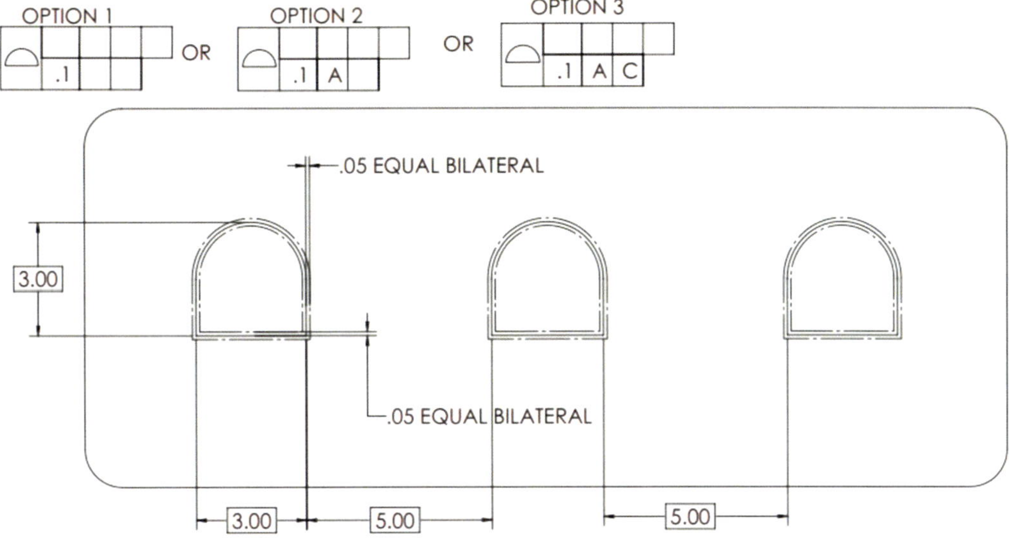

Figure 8.16: Feature Relating Tolerance Zone

8.11 No Datum is in lower segment of the Composite frame (Option 1, Figure 8.16):

If there is no datum mentioned in feature control frame, FRTZF can rotate and translate with in PLTZF. Any portion of FRTZF goes out of the boundaries of Pattern locating zone is note useful.

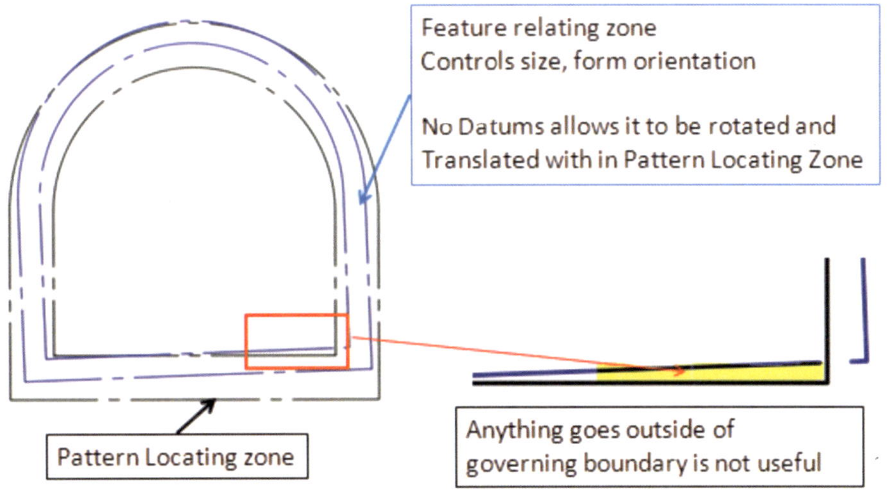

Figure 8.17: Feature Relating Tolerance Zone interpretation

8.12 Primary Datum A is repeated in lower segment of the Composite frame (Option 2, Figure 8.16):

If the primary Datum is repeated with lower segment, the FRTZF is oriented in rotation w.r.t. datum A. It can translate and rotate with other datums but within boundaries governed by PLTZF. Tolerance zone look similar as shown in 8.17 but Perpendicular to datum A is requirement.

8.13 Primary and Secondary Datum A is repeated in lower segment of the Composite frame (Option 3, Figure 8.16):

If the primary and Secondary Datum is repeated with lower segment, the FRTZF is oriented in rotation w.r.t. datum A and C. It can translate within boundaries governed by PLTZF. Tolerance zone look similar as shown in 8.17 but Perpendicular to datum A and parallel to datum C is requirement.

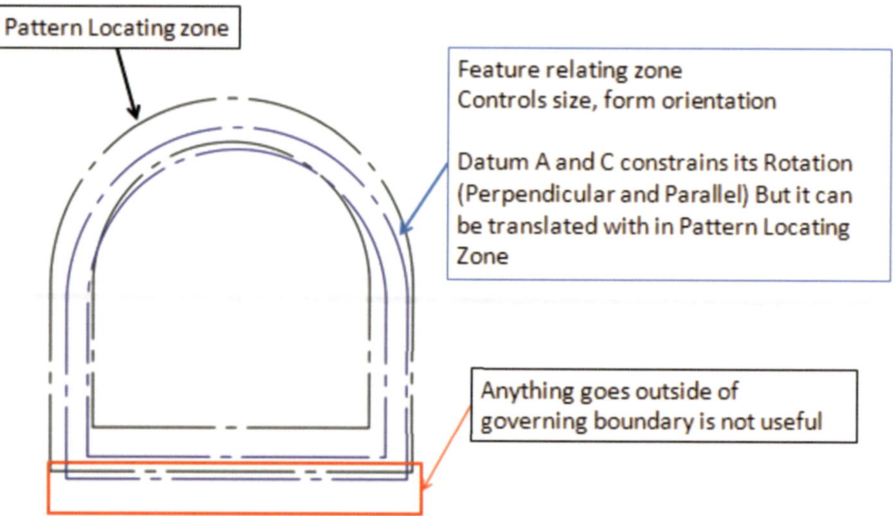

Figure 8.18: Feature Relating Tolerance Zone interpretation rotational constrain to Datum A and C

8.14 Individual feature control within Feature in Pattern:

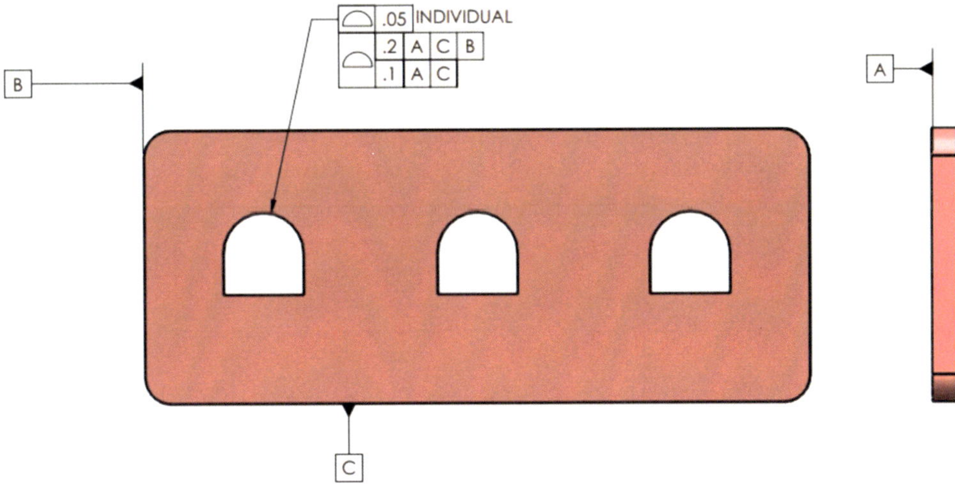

Figure 8.19: Individual application

If each profiled feature in pattern needs to be controlled for size and form individually with more tight tolerances, we use another single segment with composite profile tolerance.

The multi-segment analogy is similar to that of positional tolerancing.

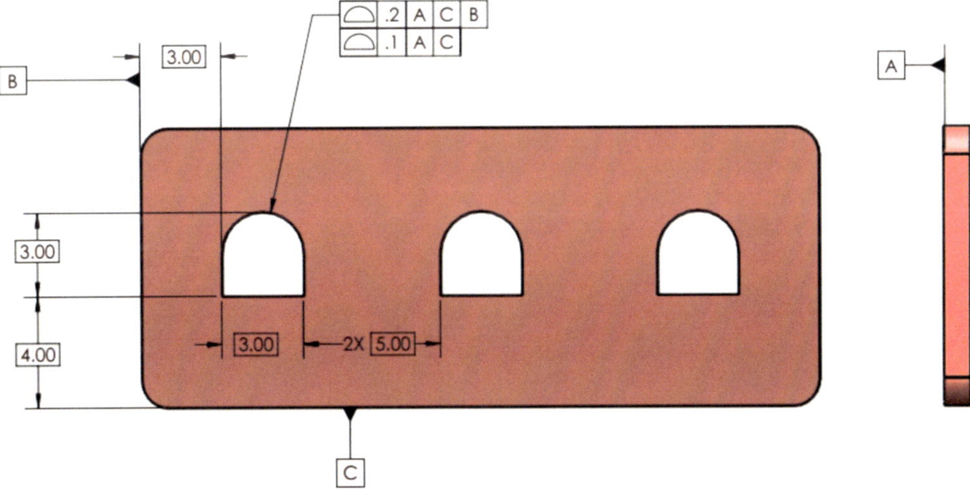

Figure 8.20: Multi-Segment

Each segment is confirmed separately. No PLTZF or FRTZF is used in multi segment application.

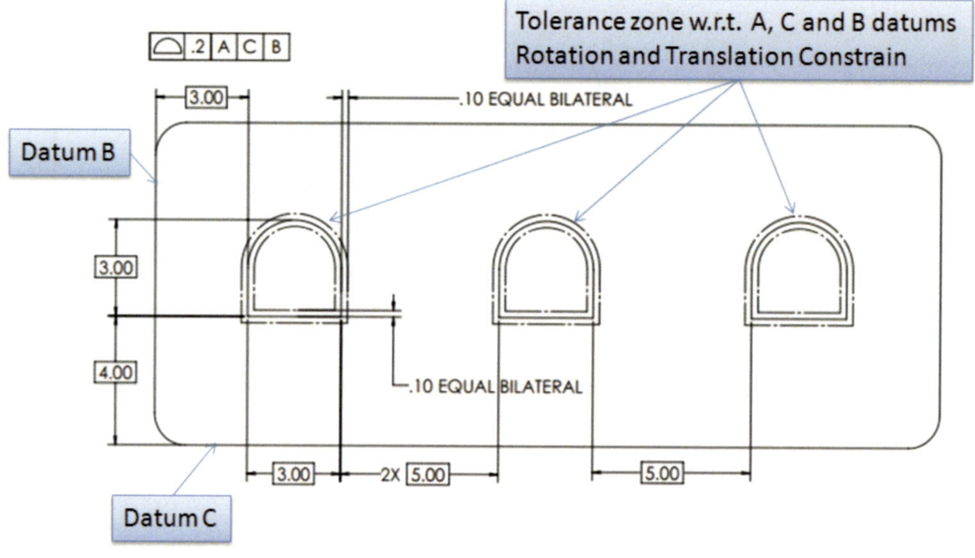

Figure 8.21: Multi-Segment – First segment w.r.t. Datum A, C and B

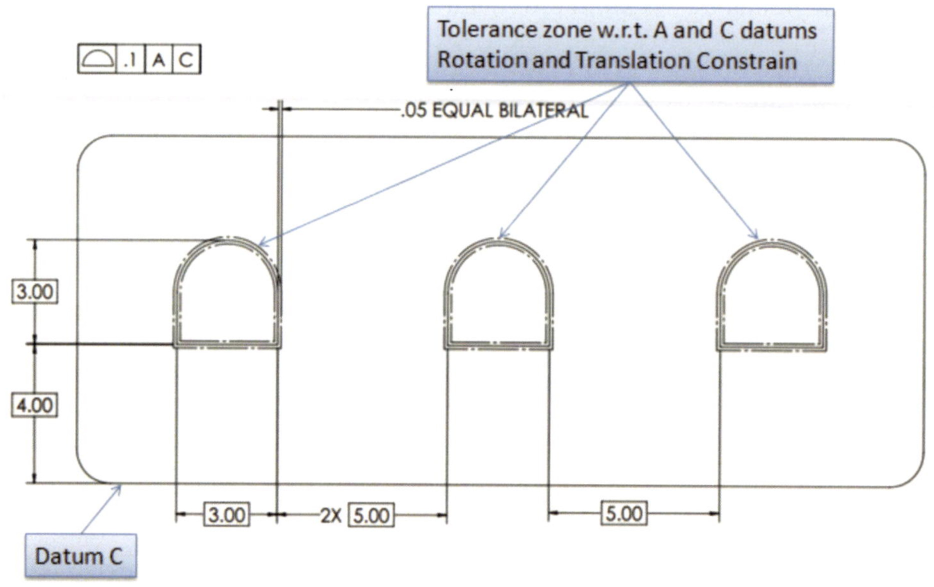

Figure 8.22: Multi-Segment – lower second segment w.r.t. Datum A, and C

If we compare figure 8.22 and 8.21 in second segment as datum B is not used, the tolerance zone can translate in right and left but with in upper segment zone. Its movement in up and down is constrained with Datum C dimension 4.00, so second segment tolerance zone cannot move up and down. As shown in figure 8.23.

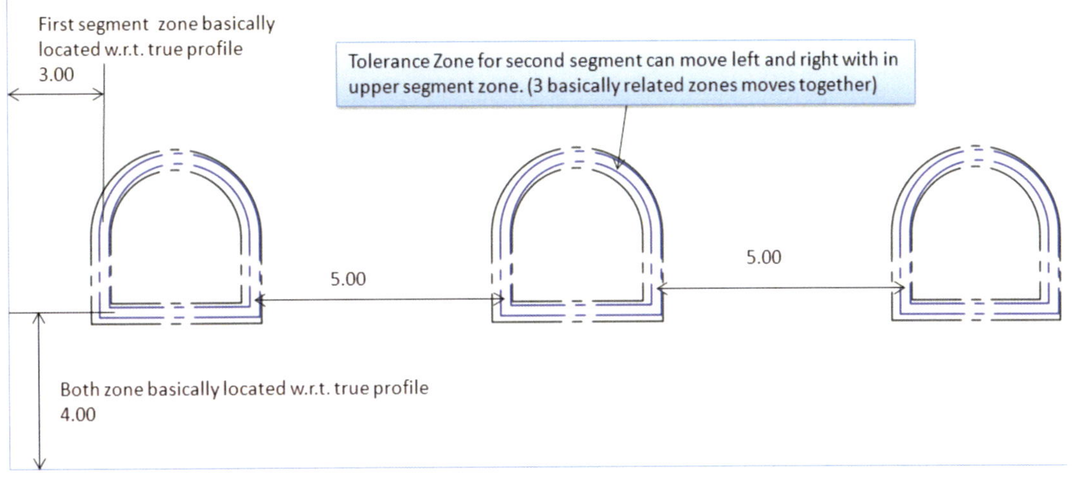

Figure 8.23: Multi-Segment

8.15 Application of Profile Tolerance:

Profile tolerance can be used in combination with Positional tolerance, Runout tolerance or orientation tolerances as refinement. It can be used to define co-planarity with or without datum. It can be used to control conicity with or without datums. If we use datum it controls size, location and form and if without datums it controls size and form.

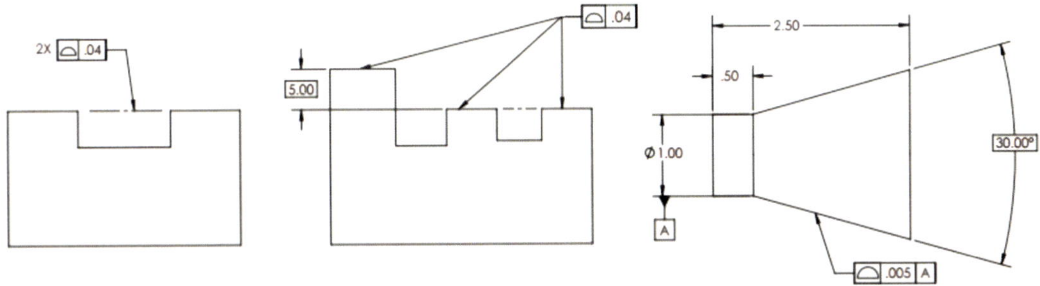

Figure 8.24: Applications

Chapter 9

Orientation Tolerance:

9.1 Orientation Tolerances:

Orientation tolerance controls parallel, perpendicular and all angular relationships. Orientation applied to flat surface controls flatness to the tolerance value, further refinement can be done by flatness tolerance itself. Orientation tolerance can be applied to feature of size as well.

Orientation tolerance does not control location of feature. Even if datum mentioned in reference frame constrained all degree of freedom, it only controls rotation for orientation tolerance. Runout, profile positional tolerances also controls orientation of controlled feature. We must consider them if applied.

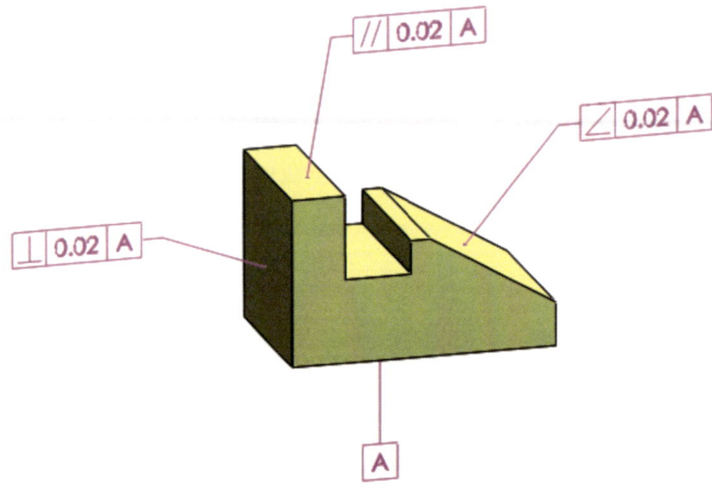

Figure 9.1: Orientation Tolerances

9.2 Angularity:

Angularity is specified as symbol shown in figure 9.1. It controls surface, center plane or axis at particular angle with specified datums.

9.3 Parallelism:

Parallelism as shown in figure 9.1. It controls surface, Center plane or axis parallel to datum i.e. all points are equidistance from Datums.

9.4 Perpendicularity:

Parallelism as shown in figure 9.1. It controls surface, Center plane or axis perpendicular (90° angle) to datum.

9.5 Tolerance Zones for Orientation Tolerances:

When Orientation tolerances applied to Flat surfaces or non circular feature of size, tolerance zone is two parallel planes with in which center plane or surface must lie. These two planes are oriented with respect to datums constrains rotational degree of freedom. Figure 9.2 tolerance zones are such tolerance zones.

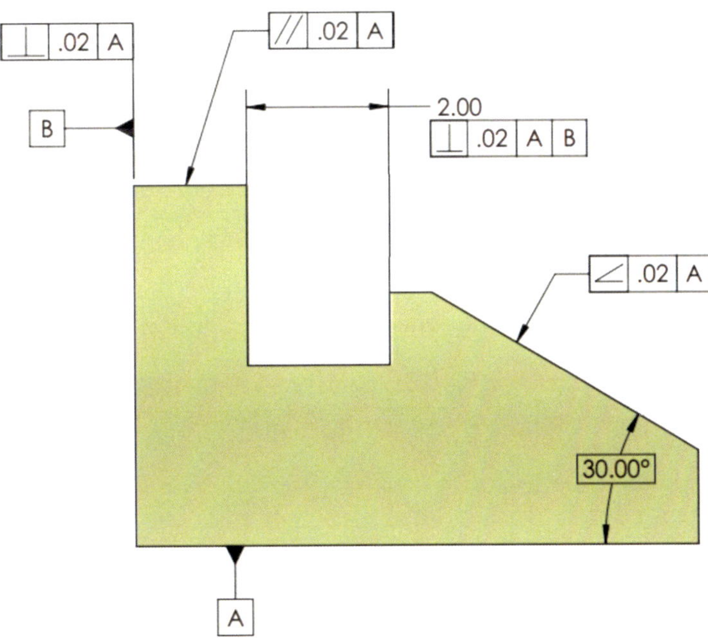

Figure 9.2: Tolerance zone - Two parallel Planes

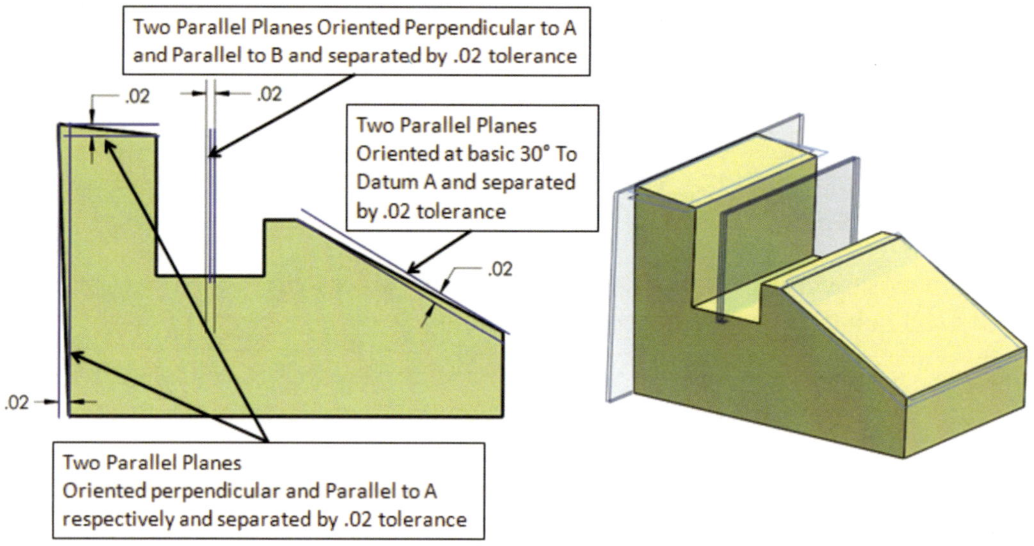

Figure 9.3: Tolerance zone - Two parallel Planes

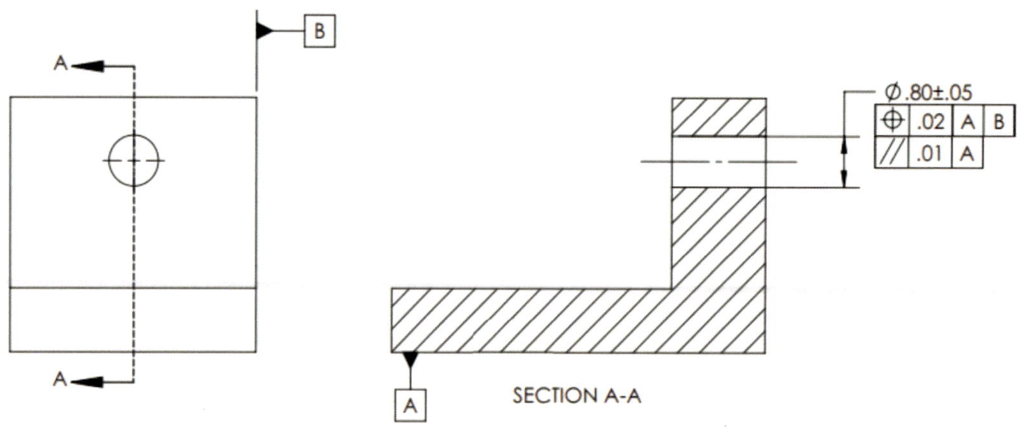

Figure 9.4: Tolerance zone - Two parallel Planes

Figure 9.4 shows tolerance zone is two parallel plane both are parallel to datum A separated by .01 within which axis should lie. This is used as refinement of positional tolerance.

Tolerance zone can be cylindrical if diameter symbol is used and orientation tolerance is applied to feature of size. The axis is parallel or perpendicular to datums as shown in figure 9.5.

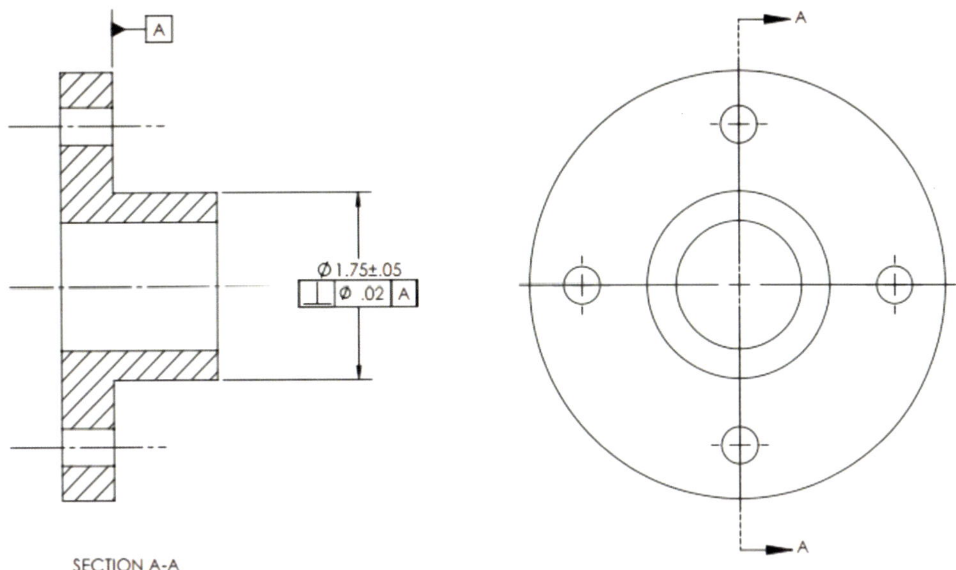

Figure 9.5: Tolerance zone – Cylinder (RFS basis)

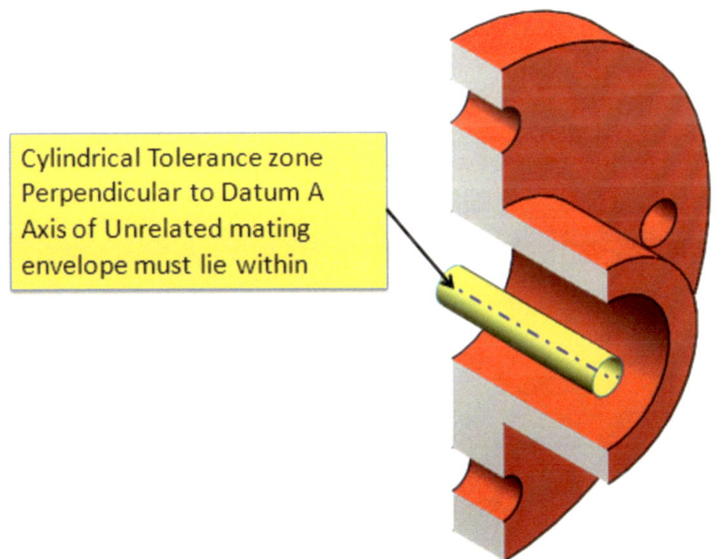

Figure 9.6: Tolerance zone – Cylinder (RFS basis)

9.6 Orientation Tolerances @ MMC modifier:

It is shown in figure 9.7 and 9.8 for Perpendicularity and Parallelism.

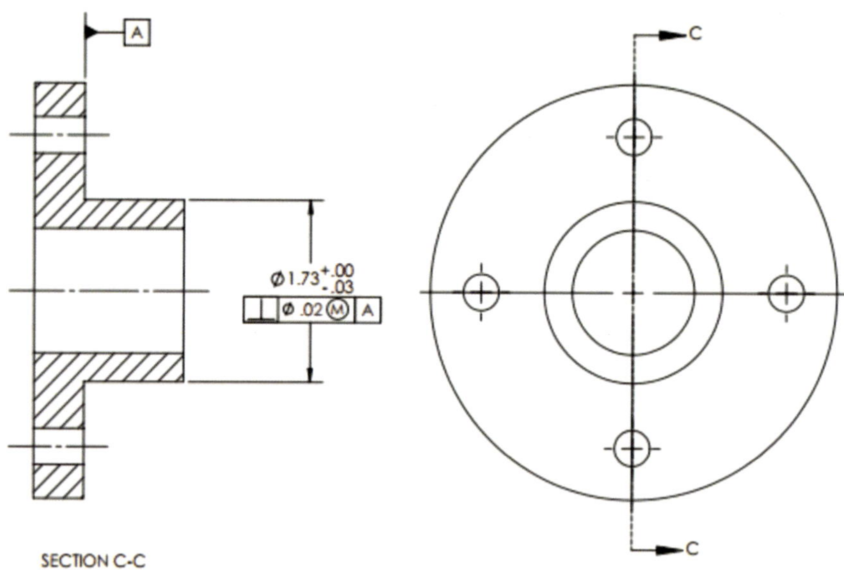

Figure 9.7: Tolerance @ MMC

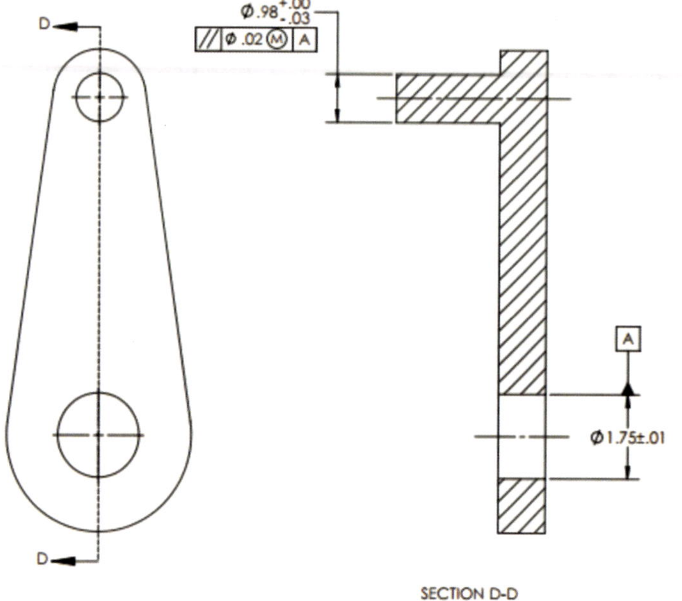

Figure 9.8: Tolerance @ MMC

If MMC modifier is used, MMC size becomes important to us apart from that if part produced at any other size with in size limit, we get bonus by amount of departure.

Surface Interpretation: If component is produced with in size tolerance i.e. unrelated actual mating envelope is with in size tolerance, A boundary (VC) oriented with respect to the datums (No location constrain) must not be violated.

For figure 9.7, VC = MMC + Geometric Tolerance (For external feature) = 1.73 +.02 = Ø1.750 Perpendicular to datum A must not be violated by actual produced surface otherwise rejected.

For Figure 9.8, VC = MMC + Geometric Tolerance (For external feature) = .98 +.02 = Ø1.00. Axis parallel with datum A must not be violated by actual produced surface otherwise rejected.

Now if we think about axis: If the part manufactured at MMC, the features axis must lie within cylindrical tolerance zone specified oriented (parallel, Perpendicular or Angular) to reference datum. The orientation of the axis of unrelated actual mating envelope must fall within the tolerance zone.

Where actual produced size departs from MMC, we get bonus tolerance. And this Bonus tolerance is equal to the amount of Departure from MMC.

Diameter	Positional Tolerance
LMC = 1.70	.05
1.71	.04
1.72	.03
MMC = 1.73	.02

Table 9.1 (Bonus tolerance for figure 7.7)

Diameter	Positional Tolerance
LMC = .95	.05
.96	.04
.97	.03
MMC = .98	.02

Table 9.2 (Bonus tolerance for figure 7.8)

As this bonus tolerance does not affect virtual condition, this do not affects functionality or interchangeability of part and it is accepted.

9.7 Zero orientation Tolerances @ MMC modifier:

It is shown in figure 9.9 and 9.10 for Perpendicularity and Parallelism.

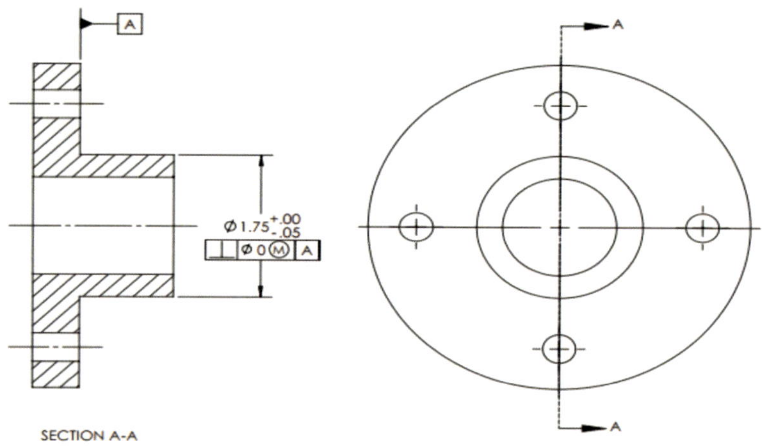

Figure 9.9: Zero Tolerance @ MMC

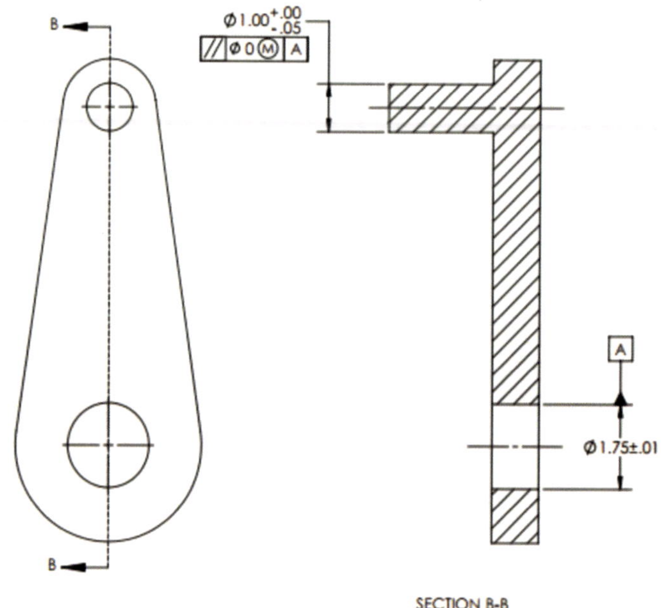

Figure 9.10: Zero Tolerance @ MMC

Whenever we use zero at MMC no orientation error permitted at features MMC size. Once feature of size departs from MMC we get orientation error by the same amount as it departed MMC. See Table 9.3.

Diameter	Positional Tolerance
LMC = 1.70	.05
1.71	.04
1.72	.03
1.73	.02
1.74	.01
MMC = 1.75	0

Table 9.3

It shows when feature size at 1.75, no orientation error permitted. Once it departs we get tolerance. Max tolerance is .05 we get that when feature at LMC. Similarly for parallelism, table 9.4 shows detail.

Diameter	Positional Tolerance
LMC = .95	.05
.96	.04
.97	.03
.98	.02
.99	.01
MMC = 1.00	0

Table 9.4

It shows Feature at mmc 1.00 will get zero orientation error. If it is manufacture at LMC it will get max tolerance i.e. .05.

Same theories go with LMC modifiers.

9.8 Projected Tolerance:

Perpendicularity can be used with projected tolerance in case of fixed fastener. Its explanation is similar to projected tolerance explained with projected tolerance. Application is shown in figure 9.11.

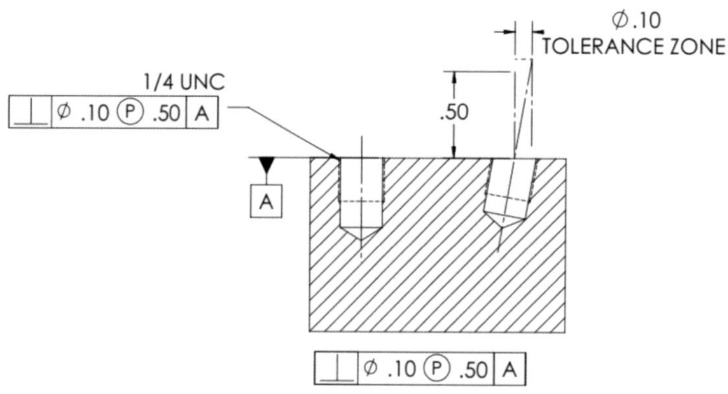

Figure 9.11: Projected Tolerance

9.9 Application of angularity:

Angularity can be used instead of perpendicularity or Parallelism for parts and it should be interpreted similarly as explained. This is all about whether we use perpendicular or parallel, ultimately is a kind of angular control. It is as shown in figure.

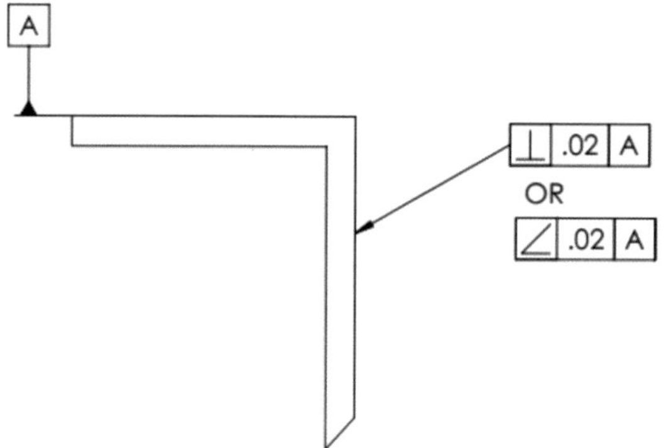

Figure 7.12: Angularity Application

Chapter 10

Runout Tolerance:

10.1 Tolerance of Runout:

Runout tolerance used to control feature revolved around an Axis or feature (surface) perpendicular to the axis. Axis should be primary datums for Runout tolerance. The datum feature used to establish axis should be at RMB and Runout tolerance does not use mmc modifiers. This is as per rule#2.

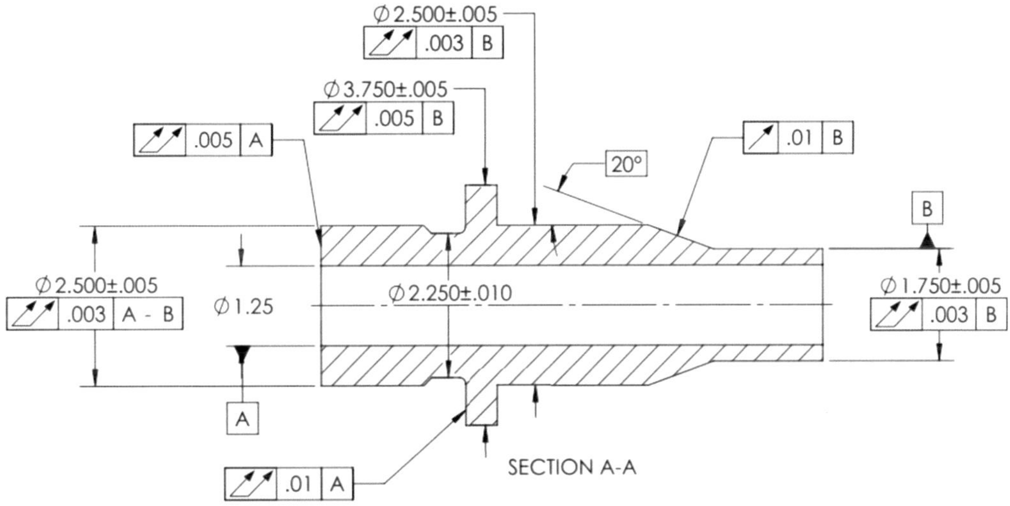

Figure 10.1: Runout Control

10.2 Datum selection for Runout:

1. Datum axis for Runout tolerance can be established by A circular feature of sufficient length.
2. Datum can be axis of feature of revolution (Primary Datum) and surface (secondary datum) perpendicular to primary datum.
3. Single datum axis can be established by multiple feature of revolution.

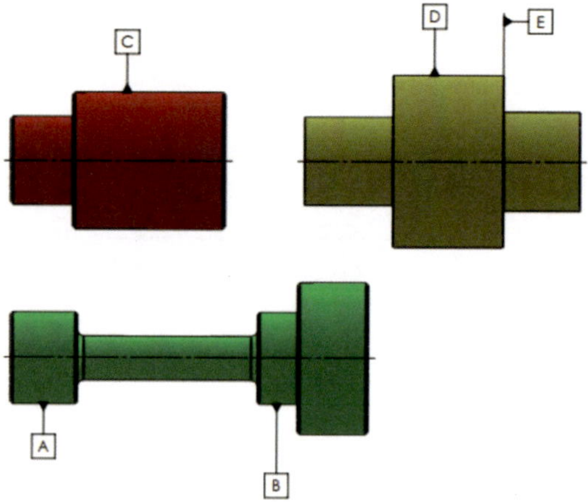

Figure 10.2: Datum Selection.

4. Runout becomes too important factor in life of a bearing (For an example). Any error in Runout will reduce the life of bearing. As the races of bearing will rapidly wear and fail if Runout error is there. So while controlling Runout, Datum selection becomes important as it should be functional one.

10.3 Circular Runout:

Circular Runout is kind of sectional property control. While controlling feature by circular Runout, it controls circularity and coaxiality with referenced datum. The tolerance zone applied individually at each cross section, the part should be rotated 360° and variation in dial gauge reading shall be within specified tolerance value.

This control even can be applied to datum itself. Whenever it is applied to surface perpendicular to datum axis, it controls wobbling.

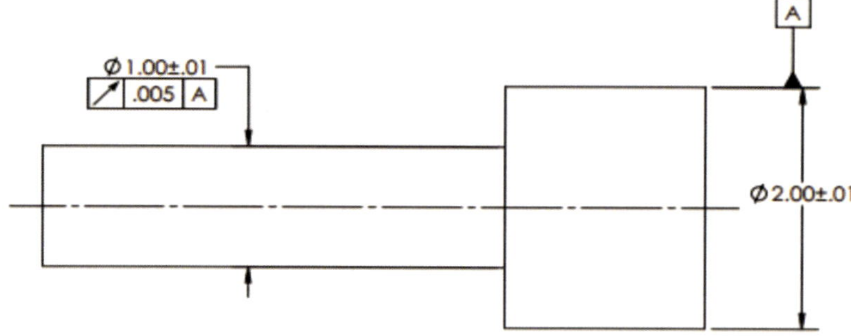

Figure 10.3: Circular Runout.

Different measurement on dial gauge as shown below:

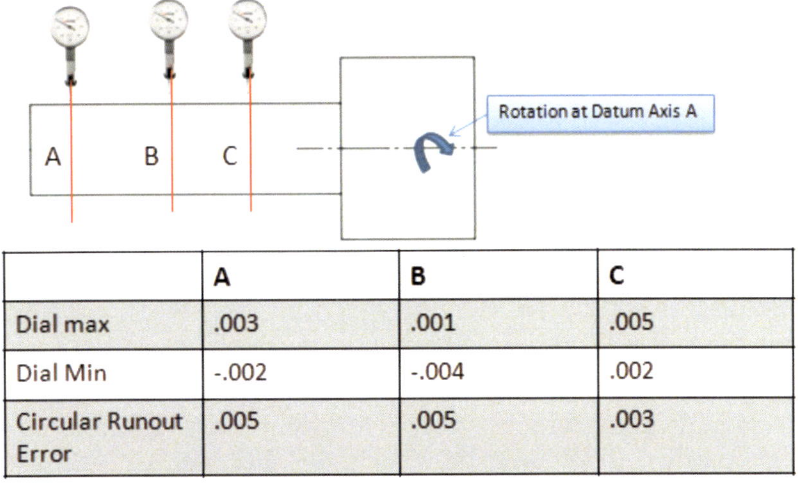

	A	B	C
Dial max	.003	.001	.005
Dial Min	-.002	-.004	.002
Circular Runout Error	.005	.005	.003

Figure 10.4: Circular Runout measurement

Dial gauge kept at different sections, each section shows error within Runout tolerance zone and part will be accepted.

10.4 Total Runout:

Total Runout controls all elements of a surface. When total Runout applied to a surface of revolution, it controls:

Circularity + straightness + Location (coaxiality) + any angularity

Let us see same example in 10.3 applied with Total Runout.

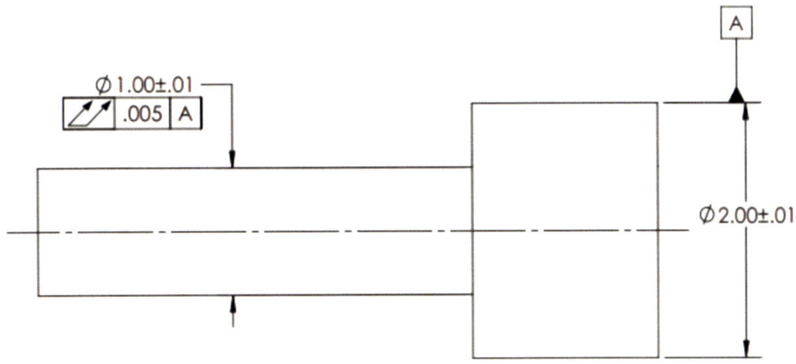

Figure 10.6: Total Runout Application

Different measurement on dial gauge as shown below:

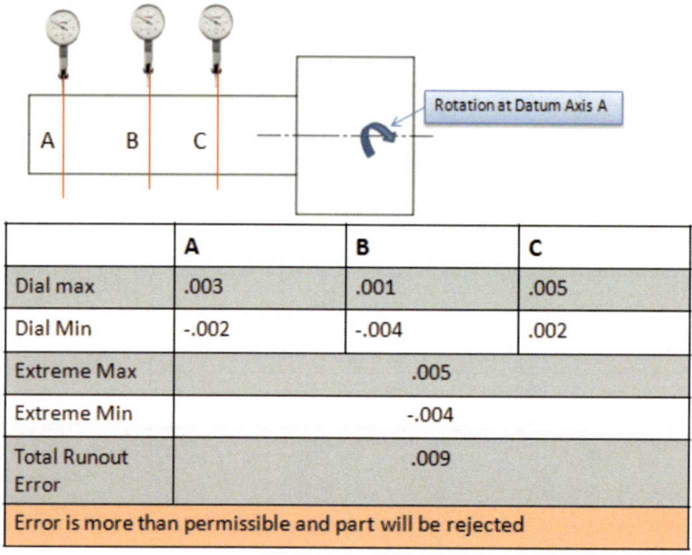

	A	B	C
Dial max	.003	.001	.005
Dial Min	-.002	-.004	.002
Extreme Max	.005		
Extreme Min	-.004		
Total Runout Error	.009		
Error is more than permissible and part will be rejected			

Figure 10.7: Total Runout Measurement

Even part is accepted with same errors and permissible value in Circular Runout application and it will get rejected in Total Runout case. It controls cumulative effect.

If total Runout applied to a surface it controls perpendicularity and Flatness.

10.5 Multiple Datum to control Runout and Control of datum surface:

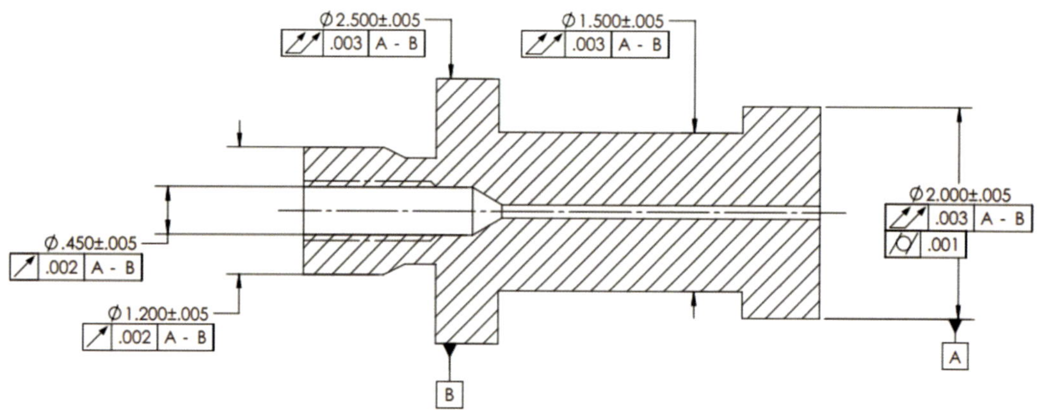

Figure 10.8: Multi datum and Individual datum control

Datum A and Datum B is simulated together to establish single datum axis. And other features are controlled with reference to datum. Datum feature itself can be controlled with Runout tolerance. Depending upon complexity single Runout may be suggested to use. Further refinement of form of datum feature may be done by using Cylindricity as shown in figure 10.8.

10.6 Flat surface as Datum:

Datum Selection is depending on functional application and design requirement. Whenever we select flat surface as datum in Runout tolerance, the surface should be at right angle to datum simulator axis. As shown in figure

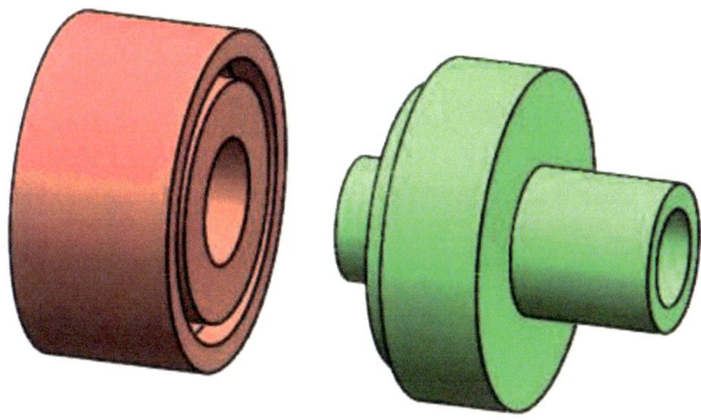

Figure 10.9: Functional Assembly

The interfacing surfaces in assembly are selected as datum.

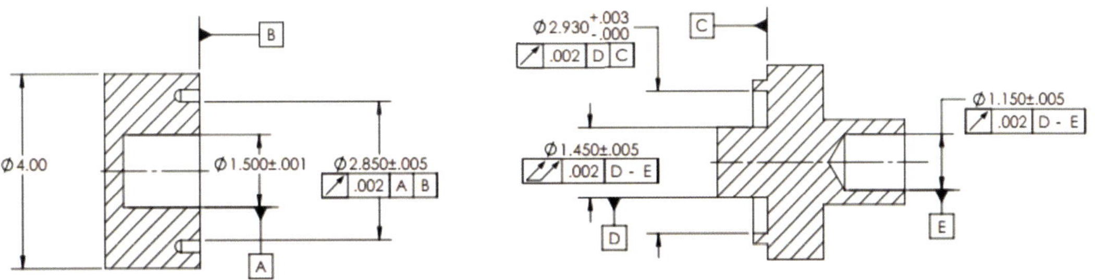

Figure 10.10: Datum Selection and Application

10.7 Application of circular Runout on a curved surface:

As we have already discussed, circular Runout is a sectional property so we can use it on curved surface of revolution. It will control circularity and coaxiality of each circular element.

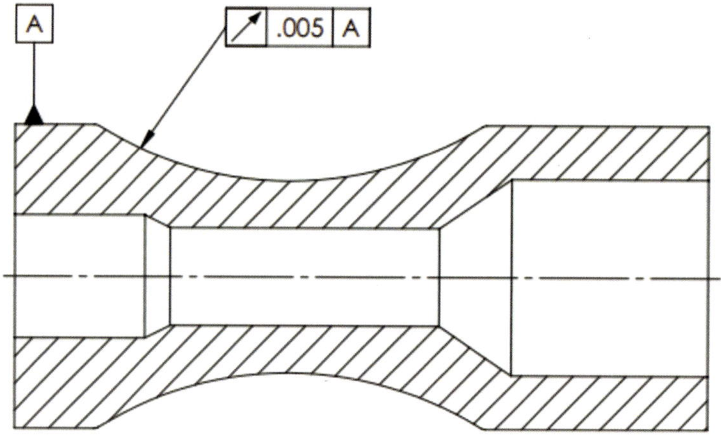

Figure 10.11: Datum Selection and Application

Appendices	Description
Appendix A	Formulas
Appendix B	ISO and ASME Y14.5

Appendix A

A.1 Resultant Condition, Virtual Condition, Inner Boundary and Outer Boundary

Boundaries	Internal Feature @ RFS	External Feature @ RFS
IB	MMC – Geometric Tolerance	LMC- Geometric Tolerance
OB	LMC + Geometric Tolerance	MMC + Geometric Tolerance
Boundaries	Internal Feature @ MMC modifier	External Feature @ MMC Modifier
VC	MMC - Geometric Tolerance	MMC + Geometric Tolerance
RC	LMC + Geometric Tolerance + Bonus	LMC - Geometric Tolerance- Bonus
Bonus	LMC-MMC	MMC - LMC
Boundaries	Internal Feature @ LMC modifier	External Feature @ LMC modifier
VC	LMC + Geometric Tolerance	LMC - Geometric Tolerance
RC	MMC - Geometric Tolerance- Bonus	MMC + Geometric Tolerance + Bonus
Bonus	LMC-MMC	MMC - LMC

A.2 Floating fastener formula:

If two plates have clearance holes and both to be assembled with Bolt we can calculate tolerance by formula as below:

$$T = H - F$$

T = Tolerance value

H = Smallest hole size

F = Major diameter of fastener (Largest Diameter)

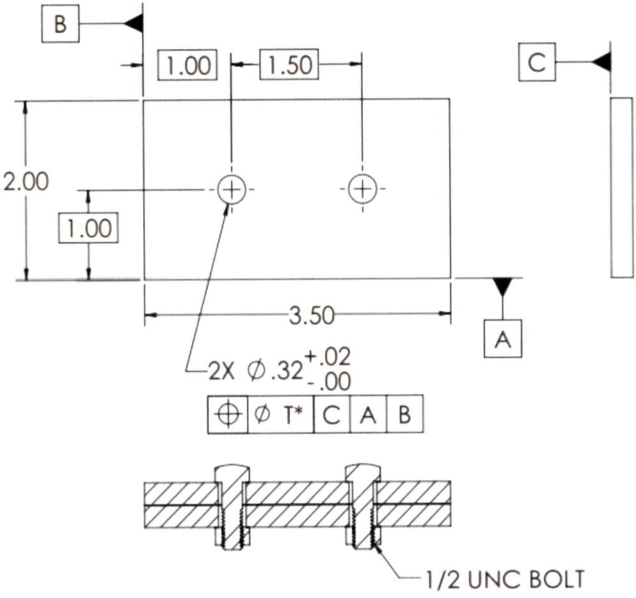

Example 1

Example 1 shows plates having clearance hole of Ø.32 and bolt size Ø.25. We need to calculate T* (Tolerance).

T = H-F = .32 - .25

= .07 (Positional tolerance)

In this case the hole both hole and fastener diameters are similar so same tolerance for both holes. If different hole and bolt sizes, we can apply T = H – F for each individually.

A.3 Fixed fastener formula:

Example 2 shows fixed assembly. It is fixed fastener assembly as one of the plates has tapped hole. We use projected type of tolerance with these applications.

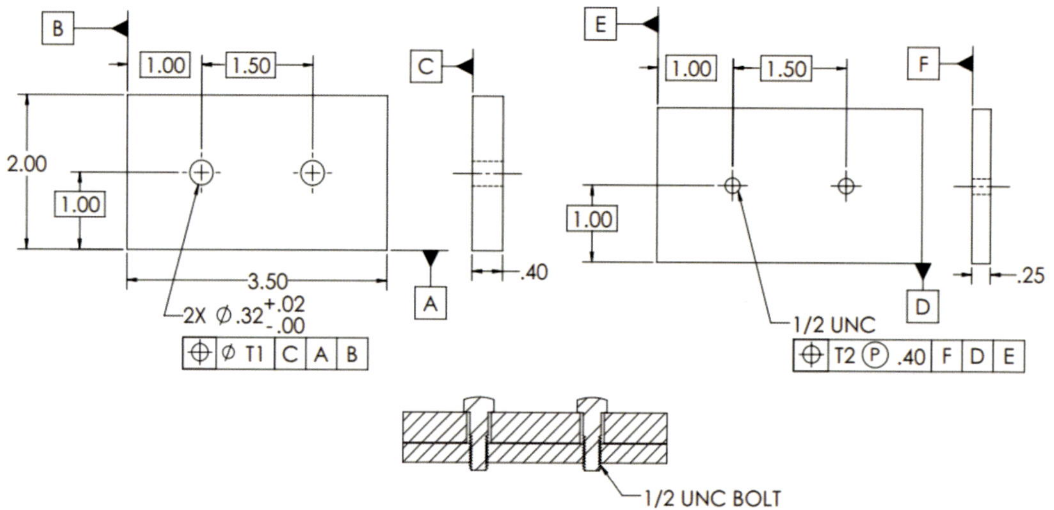

Example 2

We can use following formula to calculate T1 and T2.

$$H = D + T1 + T2$$

  H = Minimum Hole size

  D = Max Fastener diameter

  T1 = Positional tolerance on clearance hole.

  T2 = Positional Tolerance on tapped hole (Projected)

$$T1 + T2 = H - D$$

$$= .32 - .25$$

$$= .07,$$

If T1 = T2 = T

  2T = .07

  T = .035.

If T1 and T2 tolerances are different and T1 = .02. The Threaded hole can have more positional tolerance T2 = 05.

A.4 Fixed fastener formula:

When there is no projected tolerance in use we should consider the thickness of both plates that will plie important role in any assembly issue.

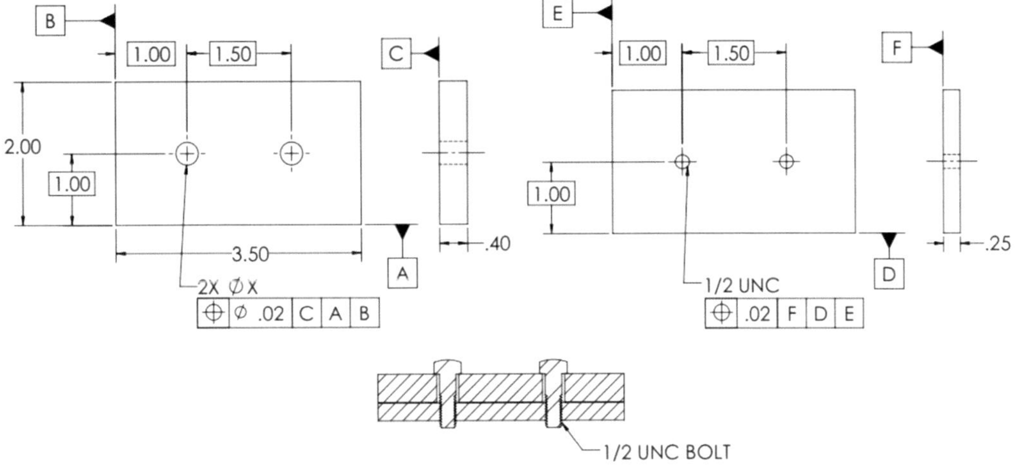

Example 3

We can use following formula to calculate X mini hole size.

$$H = X\ min = D + T1 + T2\ (1 + 2*P1\ /\ P2)$$

H = Minimum Hole size

D = Max Fastener diameter

T1 = Positional tolerance on clearance hole.

T2 = Positional Tolerance on tapped hole

P1 = Plate thickness with clearance hole

P2 = Plate thickness with tapped hole.

$$X\ min = .25 + .02 + .02\ (1 + 2*.4\ /\ .25)$$
$$= .25 + .02 + .02\ (4.2)$$
$$= .25 + .02 + .084$$
$$= .355$$
$$= Minimum\ Hole\ size$$

A.4 Coaxial Holes:

See example 4 for co-axial reference.

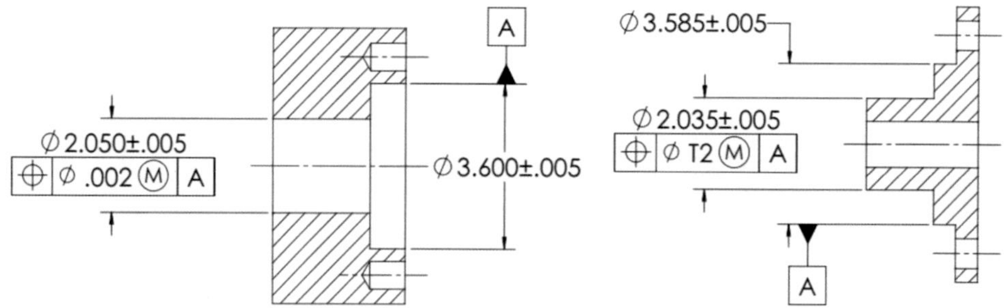

Example 4

We can use following formula to calculate T2:

H1 + H2 = D1 + D2 + T1 +T2

H1 AND H2 = Minimum Hole sizes

D1 and D2 = Max Fastener diameters

T1 = Positional tolerance on clearance hole.

T2 = Positional Tolerance on Pin diameters.

Applying to example 4

2.045 + 3.595 = 2.040 + 3.590 + .002 + T2

T2 = .008

Appendix B

B.1 ISO and ASME 2009 Y14.5 symbol comparison

ASME Y14.5	ISO	Characteristic
⎯	⎯	**Straightness**
▱	▱	**Flatness**
○	○	**Circularity**
⌭	⌭	**Cylindricity**
⌒	⌒	**Profile of a Line**
⌓	⌓	**Profile of a Surface**
∠	∠	**Angularity**
⊥	⊥	**Perpendicularity**
∥	∥	**Parallelism**
⊕	⊕	**Position**
◎	◎	**Concentricity**

≡	≡	**Symmetry**
↗	↗	**Circular Runout**
↗↗	↗↗	**Total Runout**

ASME Y14.5	ISO	Terminology
Ⓜ	Ⓜ	**Maximum Material Condition MMC**
Ⓛ	Ⓛ	**Least Material Condition LMC**
Ⓟ	Ⓟ	**Projected Tolerance Zone**
Ⓕ	Ⓕ	**Free State**
Ⓣ	**None**	**Tangent Plane**
Ⓤ	**UZ (Proposed)**	**Unequally disposed Profile Tolerance**
⟨S T⟩	**None**	**Statistical Tolerance Application**
▷	**None**	**Translation**
⌀	⌀	**Diameter**
S⌀	S⌀	**Spherical Diameter**
R	R	**Radius**
SR	SR	**Spherical Radius**

CR	None	**Controlled Radius**
( )	( )	**Reference**
A ◀ ▶ B	A ◀ ▶ B **(Proposed)**	**Between**
		**All around**
	(Proposed)	**All Over**
⊕ ⌀ .02 C A B	⊕ ⌀ .02 C A B	**Feature Control Frame**

ISO Standards for Dimensioning:

ISO/8015 Technical Drawings Fundamental Tolerance Principle

ISO/129-Technical Drawings General Principles

ISO/1101-Technical Drawings Geometrical Tolerancing

ISO/5459-Technical Drawings Datums and Datum Systems

ISO/5458-Technical Drawings Positional Tolerancing

1S0/2692-Technical Drawings Maximum Material Principle

180/3040-Technical Drawings Cones

ISO/1660-Technical Drawings Profiles

ISO/406-Technical Drawings Linear and Angular Dimensions

ISO/10578Technical Drawings Projected Tolerance Zones

ISO/2692:1988 Technical Drawings Least Material Principle

ISO/7083 Technical Drawings Symbols Proportions

ISO/10579 Technical Drawings Non-rigid Parts

ISO/2768-2-General Geometrical Tolerances

1SO/2768-1 Tolerances for Linear and Angular Dimensions